Microsoft Office Specialist
Excel Associate 365/2019
Exam Preparation

Daniel John Stine

SDC Publications
P.O. Box 1334
Mission, KS 66222
913-262-2664
www.SDCpublications.com
Publisher: Stephen Schroff

Examination Copies
Books received as examination copies are for review purposes only and may not be made available for student use. Resale of examination copies is prohibited.

Electronic Files
Any electronic files associated with this book are licensed to the original user only. These files may not be transferred to any other party.

Trademarks
Excel is a registered trademark of Microsoft, Inc. Certiport is a registered trademark of NCS Pearson, Inc. All other trademarks are trademarks of their respective holders.

The author and publisher of this book have used their best efforts in preparing this book. These efforts include the development, research and testing of the material presented. The author and publisher shall not be liable in any event for incidental or consequential damages with, or arising out of, the furnishing, performance, or use of the material.

ISBN-13: 978-1-63057-332-4
ISBN-10: 1-63057-332-9

Printed and bound in the United States of America.

Foreword

This book provides you with a collection of study materials to prepare for the Microsoft Office Specialist Excel Associate 365/2019 Exam. With a range of options for most learning styles, this text will help improve your skill level and provide an additional boost of confidence, which is sure to increase the chances of a successful exam outcome.

Study material for all learning styles, including:

 Printed book
- o **Focused Study** *on objective domains*
- o **Flashcards** *cut out with scissors*
- o **Exam Day Study Guide** *one page*

Downloads
- o **Narrated Videos** *with optional captions*
- o **Practice Software** *Prerequisite: Microsoft Excel 365/2019 installed*

Sharply focused on the required topics, the book begins with an overview of the exam process, the user interface and the five main objective domain categories, which include:

- Manage Worksheets and Workbooks
- Manage Data Cells and Ranges
- Manage Tables and Table Data
- Perform Operations by using Formulas and Functions
- Manage charts

The text concludes with an overview of the included practice exam software download. This software mimics the real exam as much as possible, in terms of user interface, number and types of questions, as well as a time constraint. While this study guide cannot claim to cover every possible question that may arise in the exam, it does help to firm up your basic knowledge to positively deal with most questions… thus, leaving more time to reflect on the more difficult questions.

Errata:
Please check the publisher's website from time to time for any errors or typos found once printed. Simply browse to www.SDCpublications.com, and then navigate to the page for this book. Click the **View/Submit errata** link in the upper right corner of the page. If you find an error, please submit it so we can correct it in the next edition.

You may contact the publisher with comments or suggestions at service@SDCpublications.com.

About the Author:

Daniel John Stine AIA, CSI, CDT, is a registered architect with over twenty years of experience in the field of architecture. He has worked on many multi-million-dollar projects, including a nearly $1 billion dollar hospital project in the Midwest. Throughout these years of professional practice, Stine has leveraged many of the Microsoft Office products to organize and manage complex projects.

He has presented internationally on architecture and design technology in the USA, Canada, Ireland, Scotland, Denmark, Slovenia, Australia and Singapore; and has been a top-rated speaker on several occasions. By invitation, in 2016, he spent a week at Autodesk's largest R&D facility in Shanghai, China, to beta test and brainstorm new features in their flagship architectural design software, Revit.

Committed to furthering the design profession, Stine teaches graduate architecture students at North Dakota State University (NDSU) and has lectured for interior design programs at NDSU, Northern Iowa State, and University of Minnesota, as well as Dunwoody's new School of Architecture in Minneapolis. As an adjunct instructor, Dan previously taught AutoCAD and Revit for twelve years at Lake Superior College. He is a member of the American Institute of Architects (AIA), Construction Specifications Institute (CSI), and Autodesk Developer Network (ADN), Autodesk Expert Elite, and is a Construction Document Technologist (issued by CSI).

In addition to Microsoft Office certification study guides, Stine has written multiple books on architectural design software, all written using Microsoft Word and published by SDC Publications.

You may contact the publisher with comments or suggestions at service@SDCpublications.com.

Many thanks go out to Stephen Schroff and SDC Publications for making this book possible!

Table of Contents

2. Manage Data Cells and Ranges

3. Manage Tables and Table Data

4. Perform Operations by using Formulas and Functions

5. Manage charts

6. Practice Exam (Provided with this Book)

6.1. Introduction

7. Study Resources

7.1.1 **Exam Day Study Guide**: one page

7.2.1 **Flashcards**: 75 cards

Index

Included Online Resources

Online resources may be downloaded from SDC Publications using access code and instructions on the inside-front cover of this book.

Practice Exam Software: Test your skills with this included resource

Videos: 70 short narrated videos covering each outcome

Notes:

0.0 Introduction

Overview

In the competitive world in which we live it is important to stand out to potential employers and prove your capabilities. One way to do this is by passing one of the Microsoft Certification Exams. A candidate who passes an exam has credentials from the makers of the software that you know how to use their software. This can help employers narrow down the list of potential interviewees when searching for candidates and reviewing resumes.

When the exam is successfully passed a certificate may be printed and displayed at your desk or included with your resume. You also have access to a Microsoft badge for use on business cards or on flyers promoting your work.

The exams <u>are</u> based on a specific release of Excel. It is important that you ensure your version of Excel matches the version covered in this book, and the version of the exam you wish to take. Ideally, you will want to take the exam for the newest version to prove you have the current skills needed in today's competitive workforce.

Important Things to Know

Here are a few big picture things you should keep in mind:

- **Practice Exam**
  - The practice exam, that comes with this book, is taken on **your own computer**
  - You need to have **Excel installed** and ready to use during the practice exam
  - You must download the practice exam software from SDC Publications
    - See inside-front cover of this book for access instructions
  - **Required files** for the practice test
    - Files are downloaded with practice exam software
    - Locate files before starting practice test
  - Note which questions you got wrong, and study those topics

- **Microsoft Office Specialist – Excel Associate 365/2019 - Exam**
  - Purchase the **exam voucher** ahead of time
    - If you buy it the day of the test, or at the test center, there may be an issue with the voucher showing up in your account
    - Note: some testing locations charge an extra proctoring fee.
  - Make a **reservation** at a test center; walk-ins are not allowed
  - A computer is provided at the test center
  - Have your Certiport **username** and **password** memorized (or written down)
  - If you fail, note which sections you had trouble with and study those topics
  - You must wait 24 hours before retaking the exam

Benefits

There are a variety of reasons and benefits to getting certified. They range from a school/employer requirement to professional development and resume building. Whatever the reason, there is really no downside to this effort.

Here are some of the benefits:

- Earn an industry-recognized credential that helps prove your skill level and can get you hired.
- Develop your skills with sample projects and exercises that emphasize real-world applications.
- Accelerate your professional development and help enhance your credibility and career success.
- Boost academic performance, prepare for the demands of a job, and open doors to career opportunities.
- Display your Microsoft certificate, use the Microsoft Certified badge, highlight your achievement and get noticed.

Certificate

When the exam is successfully passed, a certificate signed by Microsoft's CEO is issued with your name on it. This can be framed and displayed at your desk, copied and included with a resume (if appropriate) or brought to an interview (not the framed version, just a copy!).

Badging
In addition to a certificate, a badge is issued. Badging is a digital web-enabled version of your credential by **Acclaim**, which can be helpful to potential employers. This is a quick proof that you know how to use the Excel features covered by the 365/2019 Associate exam.

Certified Specialist, Expert *and* Master
While this study guide focuses solely on the **Excel Associate** exam, it is helpful to know about the other options for future consideration. There are seven different exam options. These are all paid options, not free, but when considering the value outlined previously, it is worth it. See the links at the end of this section to learn more about costs.

Associate	Expert	Master
Word	Word Expert - - - - - - - - -▸	Word Expert
Excel*	Excel Expert - - - - - - - - -▸	Excel Expert
PowerPoint -▸		PowerPoint
Access		Access *or* Outlook
Outlook		

* = Covered in this book

Microsoft Office Specialist Excel 365/2019 Associate certification
Microsoft Office Specialist (MOS) Excel Associate 365/2019 certification is an excellent way for students and professionals to validate their software skills.

- MOS Excel Associate 365/2019: **35 questions** which must be answered in **50 minutes.** *Passing: 70%*

Additional Microsoft Office certifications

In addition to the Excel Associate 365/2019 certification, which this study guide is based on, these are the other options and their format.

- PowerPoint: There are 35 questions which must be answered in 50 minutes.
- Excel Expert: There are 26 questions which must be answered in 50 minutes.
- Word Expert: There are 26 questions which must be answered in 50 minutes.
- Access: There are 31 questions which must be answered in 50 minutes.
- Outlook: There are 35 questions which must be answered in 50 minutes.

A special "Master" designation is earned if **both Expert** (Word & Excel) are passed along with the **Outlook** and **PowerPoint** _or_ **Access** exams.

All exams are live in-the-application style questions.

Exam Topics and Objectives

The MOS **Excel Associate** exam covers five main topics. The outline below lists the specific topics one needs to be familiar with to pass the test. The remainder of this book expounds upon each of these items. In fact, this is the outline for each of the remaining chapters.

1. **Manage Worksheets and Workbooks**
   - Import data into workbooks
   - Navigate within workbooks
   - Format worksheets and workbooks
   - Customize options and views
   - Configure content for collaboration

2. **Manage Data Cells and Ranges**
   - Manipulate data in worksheets
   - Format Cells and Ranges
   - Define and reference named ranges
   - Summarize data visually

3. **Manage Tables and Table Data**
   - Create and format tables
   - Modify tables
   - Filter and sort table data

4. **Perform Operations by using Formulas and Functions**
   - Insert references
   - Calculate and transform data
   - Format and modify text

5. **Manage Charts**
   - Create charts
   - Modify charts
   - Format charts

Exam Releases (including languages)

The **Certiport** website lists which languages and units of measure the exam & practice tests are available in as partially shown in the image below. For the full list, follow this link:

https://certiport.pearsonvue.com/Educator-resources/Exam-details/Exam-releases

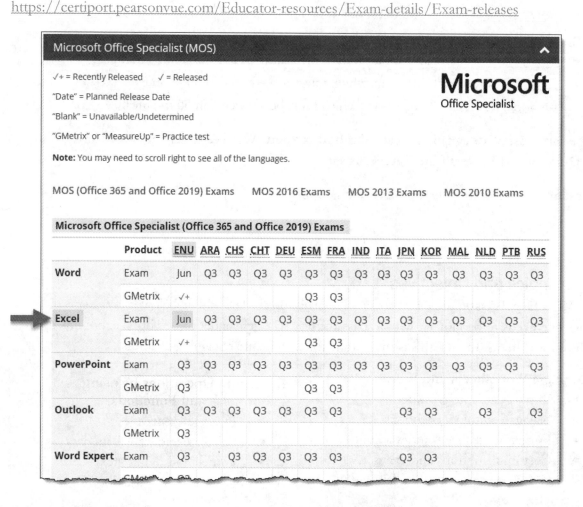

Exam releases

Certified Training Centers

There are several places to take the exam. Many academic institutions administer the exam directly to their students. Additionally, there are formal testing facilities which offer a full range of similar exams, from Yoga Instructor certification to Senior Pharmacy Technician certification.

To find the nearest testing center, start here: http://portal.certiport.com/Locator

Unfortunately, there may not be a test center in your city. For example, the closest non-academic testing facility for the author of this study guide is 150 miles away. In this case, you will have to plan a day to travel to the testing center to take the exam. In this case it is much more important to have made an appointment, purchased the voucher ahead of time and associated it with your Certiport account… and of course, studied the material well, so you do not have to retake it.

Locating a training center

From the **Certiport** frequently Asked Questions (FAQ) online page:

> "Educators and students can take the exams at a public Certiport Authorized Testing Center or become a center themselves.
>
> If schools or districts want to run exams onsite, they can easily become a testing center and run the exams seamlessly in class. Institutions can sign up to be centers on the Certiport site."

Practice Exam (included with this book)

Practice exam software is included with this book which can be downloaded from the publisher's website using the **access code** found on the inside-front cover. This is a good way to check your skills prior to taking the official exam, as the intent is to offer similar types of questions in roughly the same format as the formal exam. This practice exam is taken at home, work or school, on your own computer. You must have Excel installed, and it must be the correct version, to successfully answer the in-application questions.

This is a test drive for the exam process:

- Understanding the test software
- How to mark and return to questions
- Exam question format
- Live in-application steps
- How the results are presented at the exam conclusion

An example of the Excel practice exam is shown in the image below. When the practice exam software is started, Excel is also opened and positioned directly above. During the timed exam seven projects are presented, each consisting of a separate Excel file in which five questions must be answered by modifying the current Excel workbook. At the end, the practice exam software will grade and present the results for the exam.

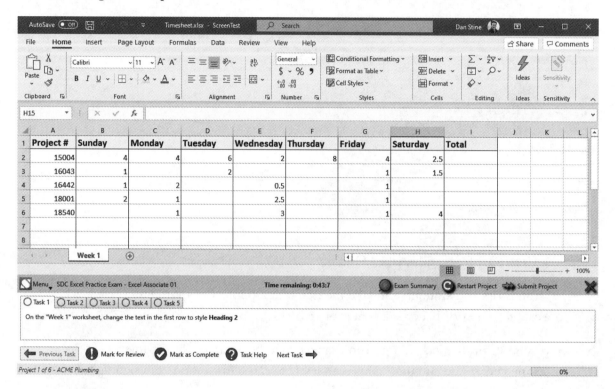

Included Practice Exam Software Example

Having taken the practice exam can remove some anxiety one may have going into an exam that may positively impact your career search.

See chapter 6 for more details on the Practice Exam software provided with this book.

Exam Preparation

Before taking the exam, you can prepare by working through **this study guide** and then the **practice exam**. You may also want to drive to the test location a day or so before the exam to make sure you know where it is and what the parking options are (if driving yourself) to ensure you are on time the day of the exam.

During the Exam

During the exam, be sure to manage your time. Quickly go through the test and answer the questions that are easy to you, skipping the ones you are not immediately sure of. The exam software allows you to view a list of questions you have not answered or have marked. Once you have answered all the easy questions you can then go back and think through those which remain. Do not exit the exam until you are completely finished, as you will not be able to re-enter the exam after that point.

> During the exam, some Excel functionality is disabled, such as Help.

Exam Results

Once the exam is finished you will receive notification of your score immediately. You must earn 700 points (out of 1000) to pass, but this is a scaled score based on weighted questions. Thus, 70% does not exactly equal a passing score. If you failed, you should note the objective areas you were not as strong in and study those areas more before taking the test again – see image below. Be sure to print your score report and take it with you to study – it is also possible to log into your Certiport account later and print it from home.

SECTION ANALYSIS	
Manage Worksheets and Workbooks	60%
Manage Data Cells and Ranges	88%
Manage Tables and Table Data	100%
Perform Operations by using Formulas and Functions	88%
Manage Charts	88%

FINAL SCORE	
Required Score	700
Your Score	892

OUTCOME	
Pass	✓

Retaking the Exam

If the exam is failed, don't worry as you can take it again – as soon as 24 hours later. If you have any doubt about your ability to easily pass the exam, consider purchasing a voucher that includes a reduced cost "retake" option.

In the event that you do not pass the exam, and you have purchased the retake option, a retake code will be emailed to you. You may re-take the exam after waiting 24 hours from the time your initial exam was first started. Retake vouchers must be used within 60 days of the failed exam.

Here is the currently posted retake policy for the certification exam:

- If a candidate does not achieve a passing score on an exam the first time, the candidate must wait 24 hours before retaking the exam.
- If a candidate does not achieve a passing score the second time, the candidate must wait 2 days (48 hours) before retaking the exam a third time.
- A two-day waiting period will be imposed for each subsequent exam retake.
- There is no annual limit on the number of attempts on the same exam.
- If a candidate achieves a passing score on an MOS exam, the candidate may take it again.
- Test results found to be in violation of this retake policy will result in the candidate not being awarded the attempted credential, regardless of score.

Resources

For more information visit these sites:

- Certiport:
  https://certiport.pearsonvue.com/Certifications/Microsoft/MOS/Overview
- Acclaim (Credly):
  https://www.youracclaim.com/
- Microsoft:
  https://www.microsoft.com/en-us/learning/certification-exam-policies.aspx

Certiport User Registration

Here are the steps to create a Certiport account, which is required to take the exam.

Start here: https://www.certiport.com/Portal/Pages/Registration.aspx

Follow the steps outlined on the site. Once complete, you will be prompted to register your account with a certification program. **Important:** be sure to select the Microsoft option in this step, and not Autodesk, Adobe, etc. This is done on the **Program tab** per the image below; just click the **Register** button (to the right of Microsoft) to get started.

Register your account with a certification program

Once you click Register, you will be prompted to verify your personal information on a new page. Once you complete this information and click Finish you will receive a confirmation email stating you are enrolled in the Microsoft certification program like the one shown below.

Congratulations on taking your first step to become Microsoft certified! »

MicrosoftOfficeCertifications@certiport.co... 7:22 AM (2 minutes ago)
to me ▾

This e-mail confirms your registration for a Microsoft Certification exam. By registering to take a Microsoft Certification exam you have joined a global community of distinguished achievers. Microsoft Certification credentials tell the world you have demonstrated proficiency in Microsoft technologies and business tools. Microsoft Certification also helps advance your career prospects by giving you a competitive edge, to increase your own personal sense of accomplishment, to establish yourself as leaders in your field, and to demonstrate your skill sets in a competitive job market.

Again, congratulations on taking that first step toward Microsoft Certification! We wish you success in all your professional and academic endeavors.

← Reply   ➡ Forward

Registration confirmation email from Certiport

User Interface

Excel is a powerful and sophisticated program. Because of its powerful feature set, it has a measurable learning curve. However, like anything, when broken down into smaller pieces, we can easily learn to harness the power of Excel.

Next, we will walk through the different aspects of the User Interface (UI). As with any program, understanding the user interface, and correct terminology, is the key to using the program's features, and using this study guide efficiently.

When Excel is first opened, the Home screen is presented, as shown in the image below. Clicking the **Blank workbook** tile (i.e. template) is the quickest way to get working in Excel. Use the **Open** option or **Recent/Pinned** files options to access existing Excel files. To verify User or Product information use the **Account** command in the lower left. Finally, the **Options** command opens the same-named dialog with a plethora of settings and options to control Excel's default behavior.

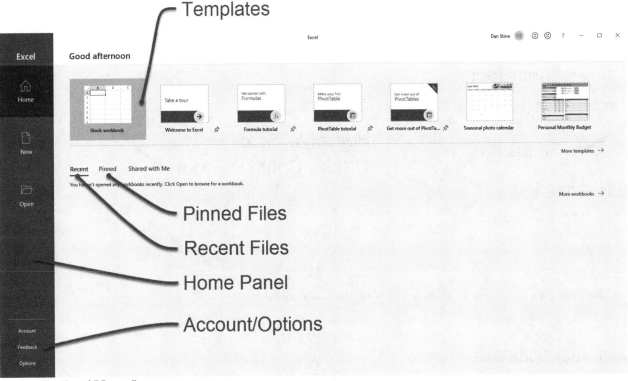

Excel Home Screen

The image below highlights important terms to know for the Excel user interface.

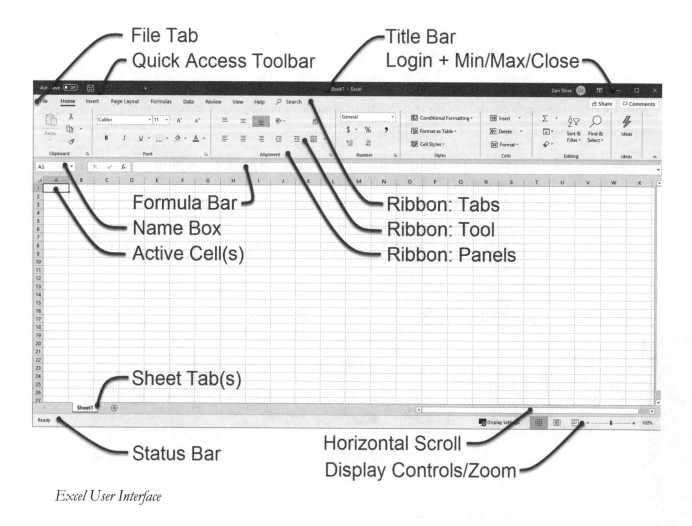

Excel User Interface

Now, let's take a closer look at some aspects of the User Interface.

Application Title Bar

In addition to the *Quick Access Toolbar* and application controls, which are covered in the next few sections, you are also presented with the product name (Excel) and the current file **name** in the center on the Application Title bar.

File Tab

Access to *File* tools such as *New, Open, Save, Save As, Share, Export* and *Print* and more. You also have access to tools which control the Excel application as a whole, not just the current workbook, such as *Options* (see the end of this section for more on *Options*).

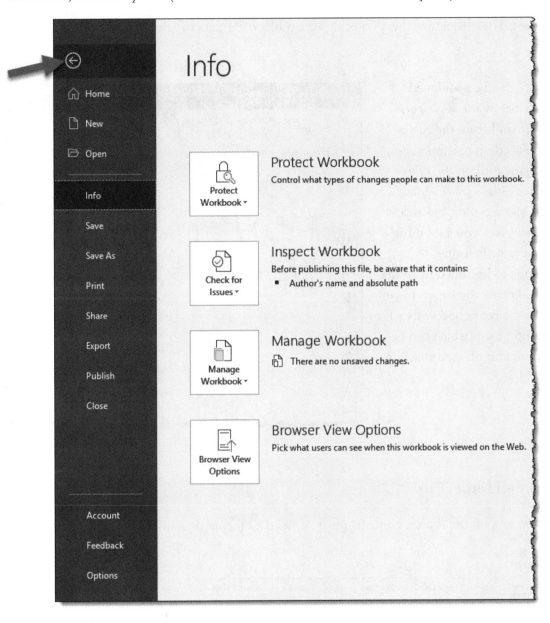

Quick Access Toolbar

Referred to as *QAT* in this book, this single toolbar provides access to often used tools: *AutoSave, Save,*

Undo, Redo. These tools are always quickly accessible, regardless of what part of the *Ribbon* is active.

The *QAT* can be positioned above or below the *Ribbon* and any command from the *Ribbon* can be placed on it; simply right-click on any tool on the *Ribbon* and select *Add to Quick Access Toolbar.* Moving the *QAT* below the *Ribbon* gives you a lot more room for your favorite commands to be added from the *Ribbon.* Clicking the larger down-arrow to the far right reveals a list of common tools which can be toggled on and off (see image to right).

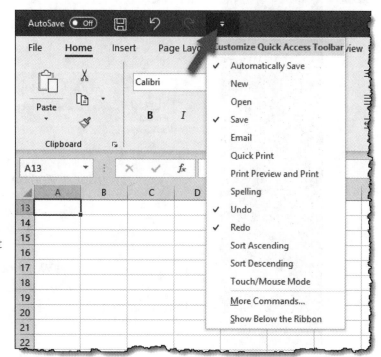

Ribbon – Home Tab

The *Home* tab, on the *Ribbon,* contains most of the data formatting and basic manipulation tools.

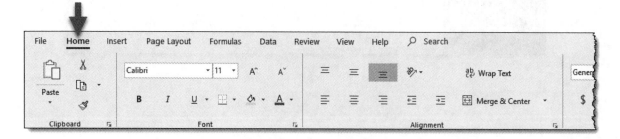

The *Ribbon* has three types of buttons: *button, drop-down button* and *split button.*

In the two following images, you can see the *AutoSum* tool is a **split button**. Most of the time you would simply click the main part of the button to Sum the selected data. Clicking the down-arrow part of the button, for the *AutoSum* tool example, gives you additional options: Sum, Average, Count Numbers, Max, Min and More Functions.

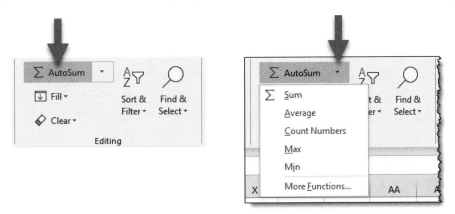

The next image is an example of a **drop-down button**. For this example, there is no dominant Fill option provided. Rather, we are required to select from a list.

Ribbon – Insert Tab

To view this tab, simply click the label "Insert" near the top of the *Ribbon*. Notice that the current tab is underlined. This tab presents a series of tools which allow you to insert objects, images and more.

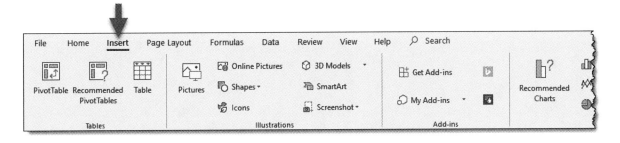

Ribbon – Page Layout Tab

The Page Layout tab controls the way data is displayed and printed.

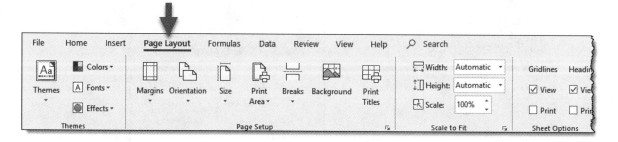

Notice that some panels have a small icon in the lower right corner; for example, the Page Setup panel as shown in the image below. Clicking this icon opens a dialog with additional related options. This small icon is officially called a Dialog Box Launcher.

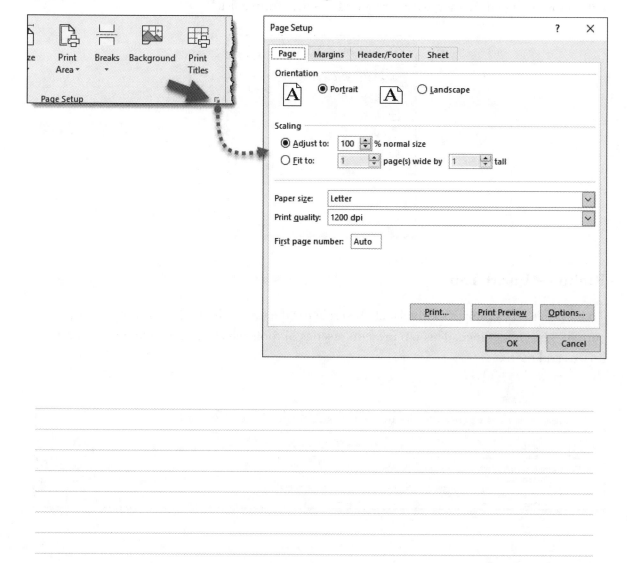

Ribbon – View Tab

The tools on the *View* tab allow you to toggle headings and gridlines on and off. Here you can also zoom and freeze portions of the worksheet, so they don't scroll off-screen.

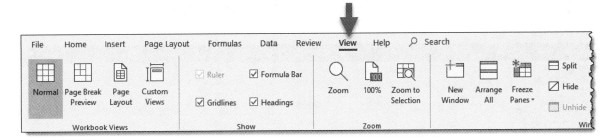

Ribbon – Add-in Tabs

If you install an **add-in** for Excel on your computer, you will likely see a new tab appear on the Ribbon. Some add-ins are free while others require a fee. The image below shows two popular PDF writer/editor tools installed: Bluebeam and Adobe Acrobat.

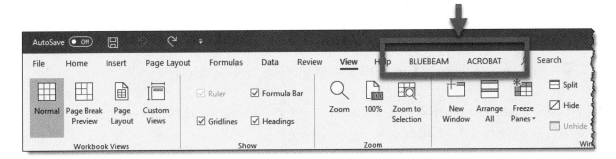

Ribbon Visibility

The *Ribbon* can be displayed in one of two states:

- Full Ribbon (default)
- Minimize to Tabs

The intent of this feature is to increase the size of the available work area, which is helpful when using a tablet or laptop with a smaller display. It is recommended, however, that you leave the *Ribbon* fully expanded

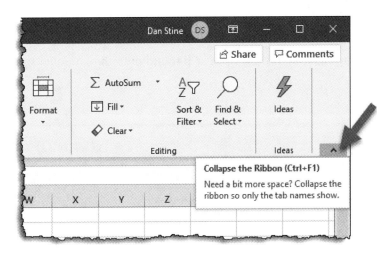

while learning to use the program. The images in this book show the fully expanded state. When using the minimized option, simply click on a Tab to temporarily reveal the tools. Click the Pin icon in the lower right to lock it open.

> Double-clicking on a Ribbon tab will also toggle the Ribbon visibility.

Minimized to Tabs

Temporarily Expanded

> The keyboard shortcut **Ctrl + F1** also toggles the Ribbon display state.

When using a tablet or touch screen, a Draw tab will also appear on the Ribbon. This is not required for the Excel Associate exam, so just ignore it.

Formula Bar

This is where dynamic and complex formulas are entered for the active cell. The checkmark (finish) and X (cancel) icons to the left are used in conjunction with formula development. The third icon is used to let Excel know you want to start a formula; this can also be done manually by clicking in the Formula Bar and entering an equal sign (=).

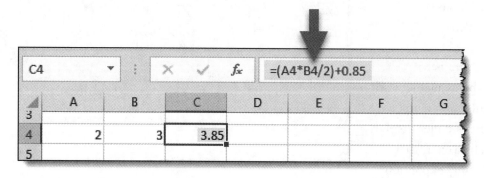

Status Bar

This area will display information about the current command on the far left (see image below). Keep your eye on this area for helpful tips while learning Excel.

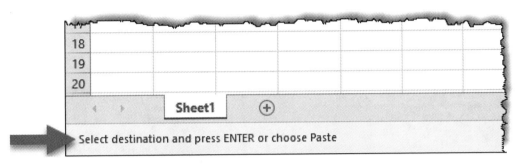

The right-hand side of the *Status Bar* shows the current zoom level and three display toggles: **Normal** (default), **Page Layout** and **Page Break Preview**. The **Display Settings** button allows users with high resolution monitors (e.g. 4k) to toggle into compatibility mode if some Excel add-ins do not support these displays. If an add-in does not support 4k screens (or Windows Text Scaling), their dialogs will be very small or parts will be jumbled and overlap, making the information hard to read.

Hover your cursor over an icon for the tool name and a brief description of what it does as shown in the image to the right. Notice, there is also a link to the Help documentation via the **Tell me more** option.

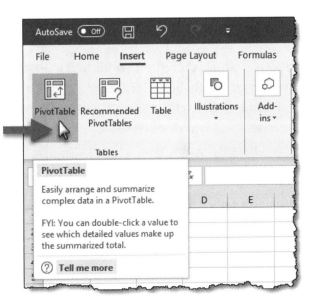

Context Menu

The *context menu* appears near the cursor whenever you right-click on the mouse. The options on that menu will vary depending on what is selected, as shown in the two examples below. Notice, the formatting toolbar also appears to facilitate quick adjustments, such as making text bold or a different color.

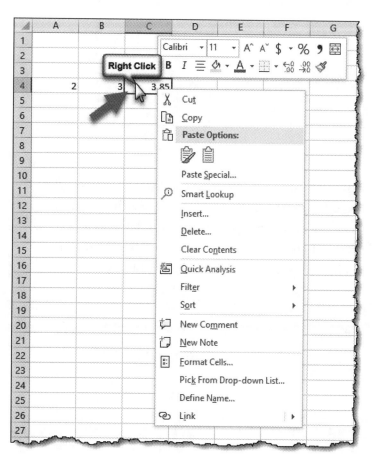

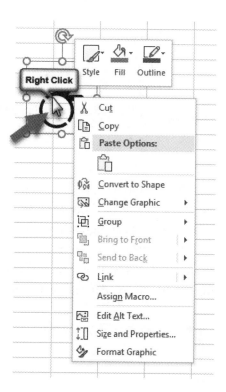

Sheet Tabs

Excel displays a tab for each sheet (worksheet) in the open Excel file. Right-click to rename, delete and more. Double-click for a quick rename option. Drag tabs to rearrange. Click the "+" icon to create a new tab.

Sheets contain distinct, but related, data. For example, a sheet for each State's tax rate.

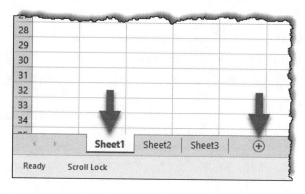

Excel Options

Accessed from the File tab, the **Options** dialog has a significant number of settings, toggles and options used to modify how the program works. It is recommended that you don't make any changes here right now. The certification exam will be based on the default settings.

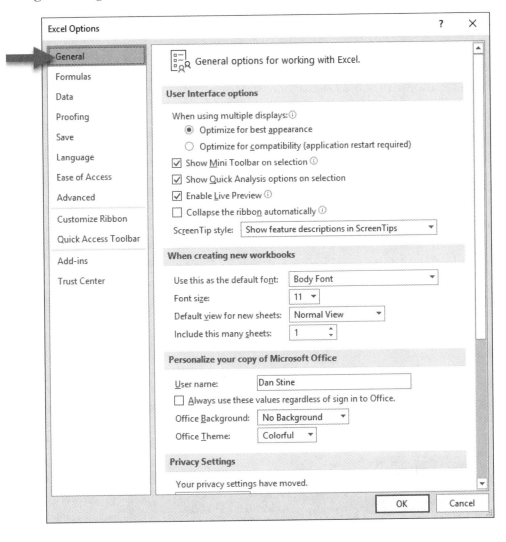

Efficient Practices

The *Ribbon* and menus are helpful when learning a program like Excel; however, many experienced users rarely use them! The process of moving the mouse to the edge of the screen to select a command and then back to where you were is very inefficient, especially for those who do this all day long, five days a week. Here are a few ways experienced Excel operators work:

- Use the **Wheel** on the mouse to scroll vertically and press and hold the wheel button while moving the mouse to scroll horizontally and/or vertically.

- Excel conforms to many of the Microsoft Windows operating system standards. Most programs, including Excel, have several standard commands that can be accessed via keyboard shortcuts. Here are a few examples (press both keys at the same time):

  - Ctrl + S   Save   *Saves the current file*
  - Ctrl + A   Select All   *Selects everything*
  - Ctrl + Z   Undo   *Undoes the previous action*
  - Ctrl + X   Cut   *Cut to Windows clipboard*
  - Ctrl + C   Copy   *Copy the selected content to the clipboard*
  - Ctrl + V   Paste   *Paste clipboard contents at cursor location*
  - Ctrl + P   Print   *Opens print dialog*
  - Ctrl + N   New   *Create new file*
  - F7   Spelling   *Launch spell check feature*

- Many Excel commands also have keyboard shortcuts. Hover your cursor over a button to see its tooltip and shortcut (if it has one). In the example shown below, press **Ctrl + Shift + =** to insert new data cells/rows/columns.

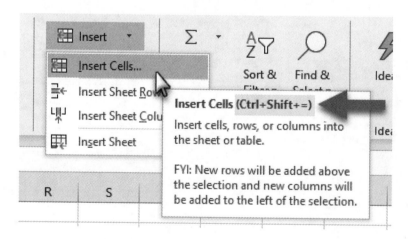

This concludes your brief overview of the Excel user interface.

1.0 Manage Worksheets & Workbooks

Introduction

In this chapter you will review some of the most basic, and essential, aspects of Excel: navigating within a workbook, import data, customize options and more.

1.1 Import data into workbooks

Review the steps required to import data from two text-based file formats: txt and csv.

1.1.1 Import data from .txt file

Many applications can export data to a text file where values are separated, or delimited, by a common character, such as a **comma** or a **tab**. The image below shows an example where values in a row are separated by tabs; also, the first tab in each row defines the first column.

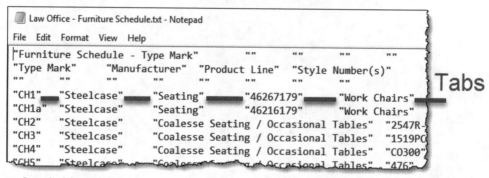

Sample delimited text file – viewed in Notepad

Excel can easily import this data into the current Worksheet.

Import data from .txt file:

1. **Get Data → From File → From Text/CVS**
2. Browse to a properly formatted text file
3. Click **Import**
4. In the import data dialog
    a. **Delimiter** – ensure proper setting, preview data will update
    b. **Data Type Detection** – set per options described in this section
5. Click **Load**

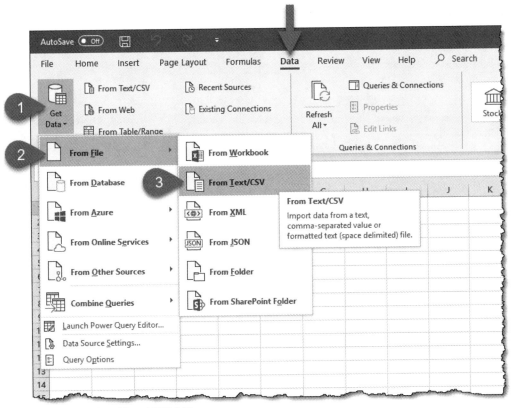

Get Data from Text/CSV file

Set the **Delimiter** value to match the character used in the .txt file. Selecting a different option causes the preview to change, which informs how the data will appear once imported. Thus, it is not necessary to open the .txt file if the character used to separate the data is not known.

Excel can scan the data in the first 200 rows, or the entire dataset, and automatically set the data type (e.g. currency, data, integer, etc.) or this can be turned off via the **Data Type Detection** list.

The **File Origin** is usually set to the correct option, as Excel scans the selected file for this.

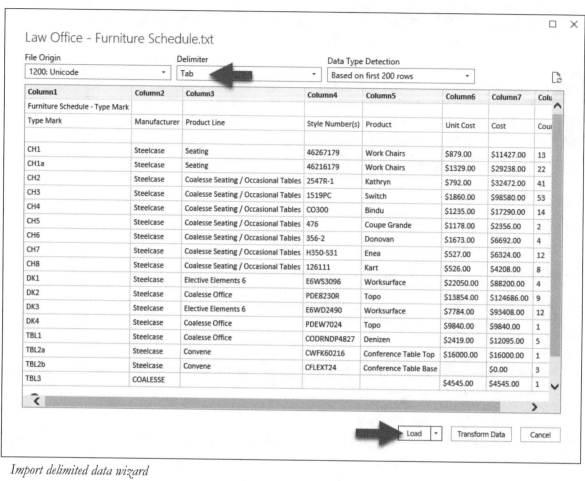

Import delimited data wizard

The result is a new table, on the current Worksheet, as shown in the image below. There is no connection to the original data source (i.e. the text file). Notice each column contains a heading with a down-arrow to quickly sort or filter the data.

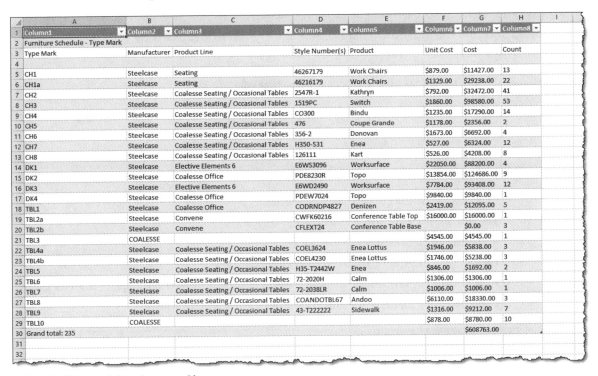

	A	B	C	D	E	F	G	H	I
1	Column1	Column2	Column3	Column4	Column5	Column6	Column7	Column8	
2	Furniture Schedule - Type Mark								
3	Type Mark	Manufacturer	Product Line	Style Number(s)	Product	Unit Cost	Cost	Count	
4									
5	CH1	Steelcase	Seating	46267179	Work Chairs	$879.00	$11427.00	13	
6	CH1a	Steelcase	Seating	46216179	Work Chairs	$1329.00	$29238.00	22	
7	CH2	Steelcase	Coalesse Seating / Occasional Tables	2547R-1	Kathryn	$792.00	$32472.00	41	
8	CH3	Steelcase	Coalesse Seating / Occasional Tables	1519PC	Switch	$1860.00	$98580.00	53	
9	CH4	Steelcase	Coalesse Seating / Occasional Tables	CO300	Bindu	$1235.00	$17290.00	14	
10	CH5	Steelcase	Coalesse Seating / Occasional Tables	476	Coupe Grande	$1178.00	$2356.00	2	
11	CH6	Steelcase	Coalesse Seating / Occasional Tables	356-2	Donovan	$1673.00	$6692.00	4	
12	CH7	Steelcase	Coalesse Seating / Occasional Tables	H350-S31	Enea	$527.00	$6324.00	12	
13	CH8	Steelcase	Coalesse Seating / Occasional Tables	126111	Kart	$526.00	$4208.00	8	
14	DK1	Steelcase	Elective Elements 6	E6WS3096	Worksurface	$22050.00	$88200.00	4	
15	DK2	Steelcase	Coalesse Office	PDE8230R	Topo	$13854.00	$124686.00	9	
16	DK3	Steelcase	Elective Elements 6	E6WD2490	Worksurface	$7784.00	$93408.00	12	
17	DK4	Steelcase	Coalesse Office	PDEW7024	Topo	$9840.00	$9840.00	1	
18	TBL1	Steelcase	Coalesse Office	CODRNDP4827	Denizen	$2419.00	$12095.00	5	
19	TBL2a	Steelcase	Convene	CWFK60216	Conference Table Top	$16000.00	$16000.00	1	
20	TBL2b	Steelcase	Convene	CFLEXT24	Conference Table Base		$0.00	3	
21	TBL3	COALESSE				$4545.00	$4545.00	1	
22	TBL4a	Steelcase	Coalesse Seating / Occasional Tables	COEL3624	Enea Lottus	$1946.00	$5838.00	3	
23	TBL4b	Steelcase	Coalesse Seating / Occasional Tables	COEL4230	Enea Lottus	$1746.00	$5238.00	3	
24	TBL5	Steelcase	Coalesse Seating / Occasional Tables	H35-T2442W	Enea	$846.00	$1692.00	2	
25	TBL6	Steelcase	Coalesse Seating / Occasional Tables	72-2020H	Calm	$1306.00	$1306.00	1	
26	TBL7	Steelcase	Coalesse Seating / Occasional Tables	72-2038LR	Calm	$1006.00	$1006.00	1	
27	TBL8	Steelcase	Coalesse Seating / Occasional Tables	COANDOTBL67	Andoo	$6110.00	$18330.00	3	
28	TBL9	Steelcase	Coalesse Seating / Occasional Tables	43-T222222	Sidewalk	$1316.00	$9212.00	7	
29	TBL10	COALESSE				$878.00	$8780.00	10	
30	Grand total: 235						$608763.00		
31									
32									

Result of imported data from text file

A .txt and .csv file can also be opened directly with Excel. In the **Open** dialog, change the **Files of Type** list to **Text Files (*.prn, *.txt. *.cvs)** as shown below. Once opened, the file may be edited in the current format or saved as a .xlsx Excel workbook.

Changing 'Files of Type' in the Open dialog

1.1.2 Import data from .csv file

The steps to import a .csv file are the same as importing a .txt file just covered. They are both text-based file formats with data delimited, i.e. separated, by a specific character.

1.2 Navigate within workbooks

Workbooks and worksheets can become large and complex, so it is important to know how to move around and find information. That is what this section of the exam focuses on.

1.2.1 Search for data within a workbook

Because workbooks, and individual worksheets, can contain a lot of data, you may need to use the Find command to locate information quickly.

Steps to search for data:

1. **Home → Find & Replace → Find…** *or* **Ctrl + F**
2. Type what you are looking for, for example: **office**
3. Click one of the 'find' buttons
   a. **Find Next**: highlight the first instance
   b. **Find Next** (again): find the next instance, repeat until the desired data is located
   c. **Find All**: lists all instances along with location information

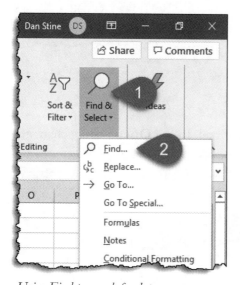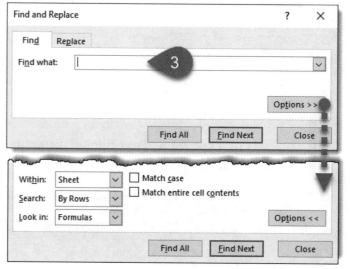

Using Find to search for data

Additional Options:

1. **Within:** Sheet or workbook
2. **Search:** By Rows or By Columns
3. **Look in:** Formulas, Values, Notes or Comments
4. **Match case:** results must match as typed, if checked
5. **Match entire cell contents:** no partial matches allowed, if checked
6. **Format** (*not shown*): narrow search results based on format: font, fill, alignment, etc.

The image below shows 'Find' and 'Find All' results for "Office" in a sample workbook.

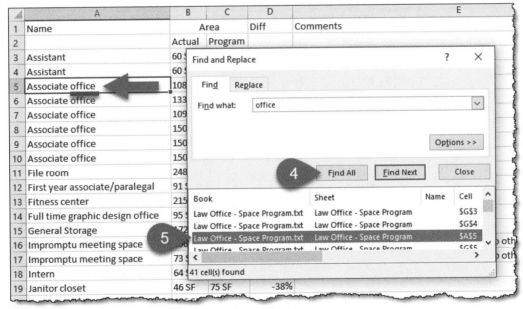

Results of using Find and Find All in a sample workbook

Advanced Search Options: *Select all cells which contain…*

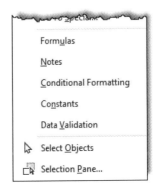

1. **Formulas:** *…a formula to derive a value.*
2. **Notes:** *…a note or comment.*
3. **Conditional Formatting:** *…conditional formatting.*
4. **Constraints:** *…constraints.*
5. **Data Validation:** *…data validation.*

1.2.2 Navigate to named cells, ranges, or workbook elements

Where the previously covered Find common looks for data based on cell contents, the Name box and Go To commands look for cells based on their location in the current sheet.

To navigate to a cell, range or named location:

In the **Name** box, do one of the following:
1. **Enter** a cell location; e.g. A4
2. **Enter** a cell range; e.g. A4:A8
3. **Select** a named item from the list
4. *Or* **Ctrl + G** → **Special** (button) to select Last Cell, Blank Cells, etc.

> During the exam, type in the given range to reduce errors from manually selecting.

Selecting cells, ranges and named locations

1.2.3 Insert and remove hyperlinks

Another way to efficiently navigate within an Excel workbook is by adding links, aka hyperlinks, to cells and objects, to instantly jump to another area or sheet. And, as it is often not possible to contain all needed data within a workbook, links can open external files and web pages.

Link to **Existing File or Web Page**

1. Select the text, cell or object
2. **Click Insert (tab) → Link** (*or* right-click → Link, *or* Ctrl + K)
3. Enter an address:
   a. URL address, e.g. https://www.steelcase.com/
   b. File path: e.g. c:\Resources\Tables.pdf
   c. *Or,* click browse for a file/web page via icons in upper right
   d. *Optional:* Click the **ScreenTip…** button and enter a tooltip message
4. Click **OK**
5. Single-click new link to access referenced data, file or web page.

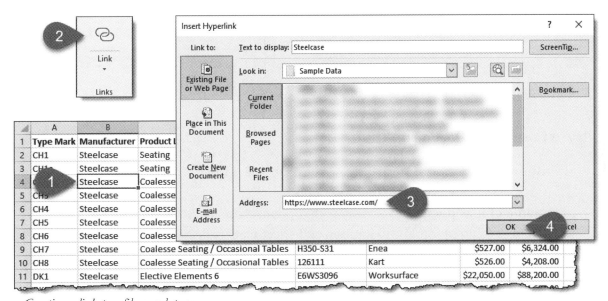

Creating a link to a file or web page

Like most hyperlinks on web pages, the formatting of the text has changed to 'underlined and blue' as shown to the right. Hover over the text to see a tooltip indicating the hyperlink reference. Now, a single-click opens the link and a 'click and hold' selects the cell or object.

Resultant link to external data

Link to **Place in This Document**

1. **Right-clic**k on desired text, cell or object and select **Link**
2. Click the **Place in this Document** option
3. **Type the cell reference** e.g. A1 or A1:B3
4. **Select worksheet** e.g. Steelcase
5. Click **OK**

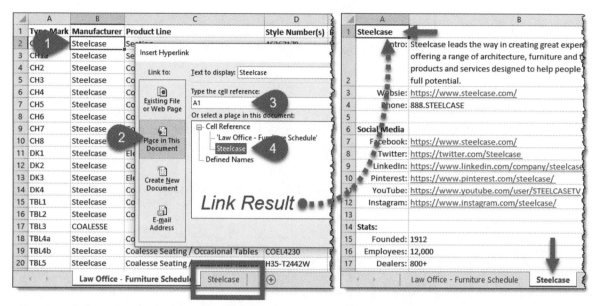

Creating a link to a location in the current document   *Resultant link to place in current document*

Link to **Create New Document**

1. **Right-click** on desired text, cell or object and select **Link**
2. Click the **Create New Document** option
3. **Name of new document**: specify path (optional here) and new file name
4. **Change** (button): browse to location and select file type; e.g. Word, Excel, etc.
5. **When to edit**: select to edit the new file now or later

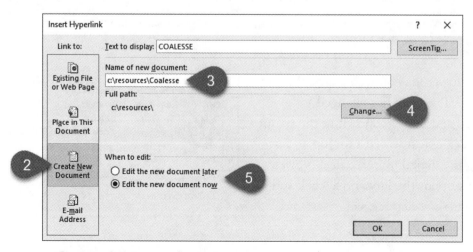

Creating a link to a new document

Link to **E-mail Address**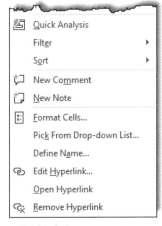

1. **Right-clic**k on desired text, cell or object and select **Link**
2. Click the **E-mail Address** option
3. **E-mail address:** enter a valid email address
4. **Subject**: enter a valid subject for the e-mail, e.g. Furniture Selection
5. **Recently used e-mail addresses:** select from list to re-use an e-mail address

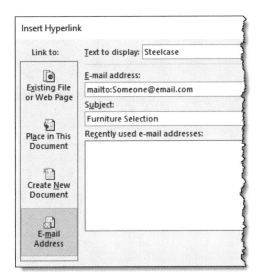

Creating a link to an email address

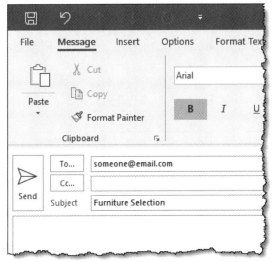

Resultant link opens Outlook with to/subject entered

Edit a Link

1. **Right-click** on desired text, cell or object and select **Link**
2. Click **Edit Hyperlink…** (see image to right)
3. Modify per steps previously outlined

Remove a Link

1. **Right-clic**k on desired text, cell or object and select **Link**
2. Click **Remove Hyperlink** (see image to right)

Right-click menu

1.3 Format worksheets and workbooks

Adjusting the way worksheets appear can be very helpful in finding the right data quickly. Especially if the workbook is being used by people who did not create it.

1.3.1 Modify page setup

The Page Setup dialog has an array of options used to control how a worksheet prints to PDF or hardcopy.

This dialog has four tabs, as follows.

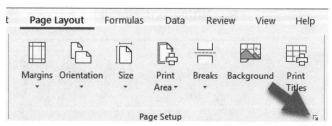

Page Setup dialog launcher

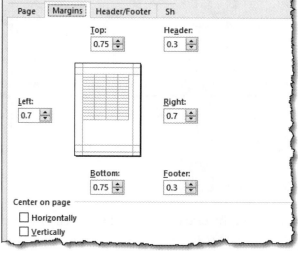

Page Setup dialog – Page and Margins tabs

Page tab

1. Orientation
   a. **Portrait**: printed page viewed as a taller document
   b. **Landscape**: printed page viewed as a wider document
2. Scaling
   a. **Adjust to**: a percentage of the normal, or original, size
   b. **Fit to**: create a poster spread across multiple pages
3. **Paper size**: predefined paper sizes, e.g. letter (8.5x11), tabloid (11x17), legal (8.5x14)
4. **Print quality**: Higher dots per inch (dpi) improves print quality on supported printers
5. **First page number**: Auto to start at 1, or manually enter a number for the first page

Margins tab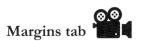

1. Margins
   a. **Top, Right, Bottom, Left**: distance from data to edge of page
   b. **Header, Footer**: distance between header/footer and top/bottom of page
2. Center on page
   a. **Horizontally**: center the data, i.e. defined print area, on the page, within margins
   b. **Vertically**: center the data, i.e. defined print area, on the page, within margins

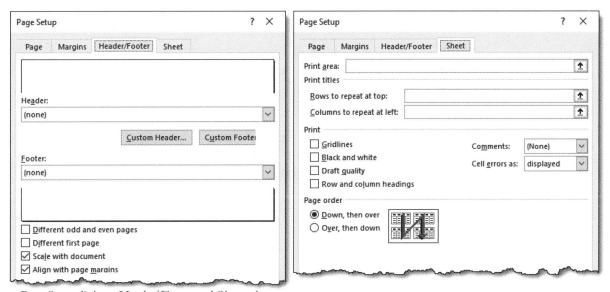

Page Setup dialog – Header/Footer and Sheet tabs

Header/Footer tab

1. **Header**: select a built-in header option from the list
2. **Custom Header…**: create a custom header in the Header dialog
3. **Custom Footer…**: create a custom footer in the Footer dialog
4. **Footer**: select a built-in footer option from the list
5. Options
   a. **Different odd and even pages**: define two header/footer (like this book)
   b. **Different first page**: define a unique header/footer for the first page
   c. **Scale with document**: scale font size of header/footer with the document
   d. **Align with page margins**: ensures header/footer always aligns with margins

Sheet tab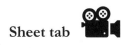

1. **Print Area**: enter a range or click the **up arrow** icon to select (via click and drag) within the document, then click the **down arrow** icon to restore the Page Setup dialog
2. Print titles
    a. **Rows to repeat at top**: repeat the selected row(s) on each page, think "titles"
    b. **Columns to repeat at left**: repeat the selected column(s) on each page
3. Print
    a. **Gridlines**: print the lines shown between cells, defining rows and columns
    b. **Black and white**: creates a black and white print from document containing color. Helpful when creating a PDF, but not required for B/W only printers.
    c. **Draft quality**: faster draft prints if supported by printer
    d. **Row and column headings**: check to include the letters and numbers
    e. **Comments**: options are None, At end of sheet, or As displayed on sheet
    f. **Cell errors as**: select how errors should appear, displayed, blank, --, n/a
4. Page order (when data does not fit on a single page)
    a. **Down, then over**: refer to the graphic in the dialog
    b. **Over, then down**: refer to the graphic in the dialog

1.3.2 Adjust row height and column width

The exam requires you to know how to change the height of a row and the width of a column.

Adjust row height 🎥

1. **Right-click** on a row header, *or multiple previously selected rows*
2. Select **Row Height…**
3. Enter new **Row height value**

Adjust column width 🎥

1. **Right-click** on a column header, *or multiple previously selected columns*
2. Select **Column Width…**
3. Enter new **Column width value**

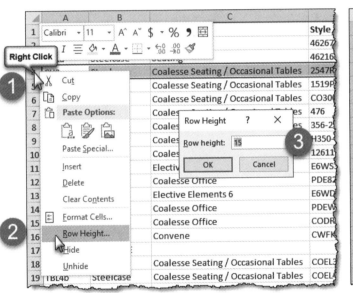

Adjust row height

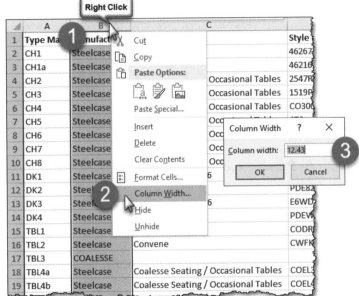

Adjust column width

In addition to changing the row height and column width manually, there are related options via **Home → Format** as shown to the right. The two **AutoFit** options fit the row/columns to the current data. The **Default Width…** matches the original values when the workbook was created.

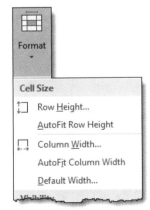

Format list (Home tab)

1.3.3 Customize headers and footers

Include predefined or custom header and footings when printing. For additional information on this topic, see the section on Page Setup previously covered.

Insert a header (similar steps for footers)

Access Page Setup dialog

1. **Page Layout** *tab* → **Page Setup** dialog launcher
2. Select the **Header/Footer** tab
3. Select from the Header list for built-in options

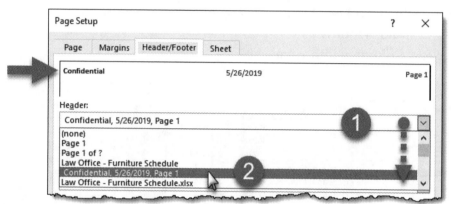

Insert built-in header

Insert a custom header (similar steps for footers)

1. **Page Layout** *tab* → **Page Setup** dialog launcher
2. Select the **Header/Footer** tab
3. Select the **Custom Header...** button
4. Click in a location, add spaces if needed, and then click icons to insert fields

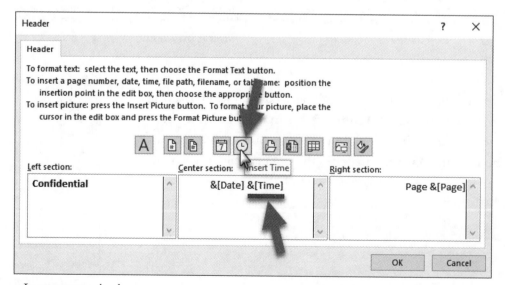

Insert a custom header

1.4 Customize options and views

This section reviews how information is displayed, or hidden, to improve the look and efficiencies related to working with data in Excel.

1.4.1 Customize the Quick Access toolbar

The Quick Access toolbar (QAT) provides access to often used tools: AutoSave, *Save*, *Undo*, *Redo*. These tools are always quickly accessible, regardless of what part of the *Ribbon* is active.

The QAT can be positioned above or below the Ribbon and any command from the Ribbon can be placed on it; simply right-click on a tool and select **Add to Quick Access Toolbar**.

Moving the QAT below the Ribbon provides more room for favorite commands to be added. Clicking the larger down-arrow to the far right reveals a list of common tools which can be toggled on and off (see image to right). Select **More Commands…** to access the Options dialog shown below, which provides a list of nearly all Excel commands.

Quick Access Toolbar options list

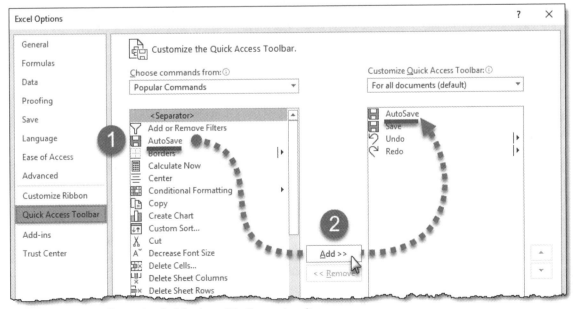

Options dialog – Customize Quick Access Toolbar options list

1.4.2 Display and Modify workbook content in different views

There are three default toggles to control the view options; they are on the **View tab** and right side of the status bar.

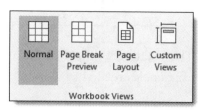

- **Normal**: default view showing all rows/columns
- **Page Break Preview**: Shows manual/automatic breaks
- **Page Layout**: based on cumulative page setup options

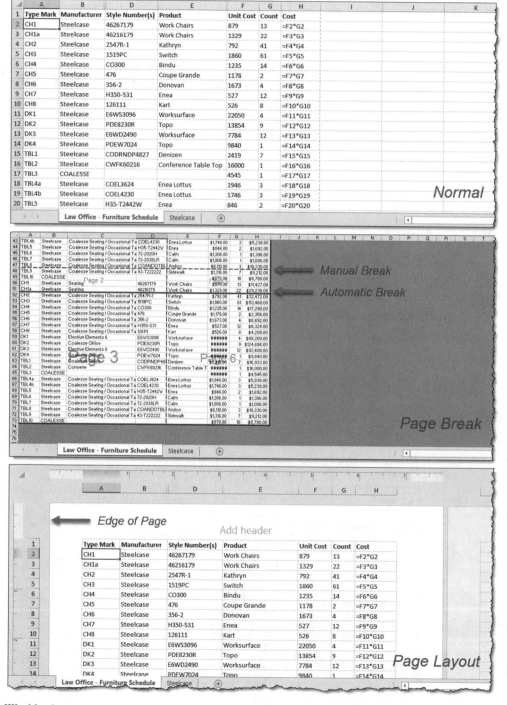

Workbook view options

1.4.3 Freeze Worksheet rows and columns

To keep track of data types and names, it is helpful to freeze the first row and/or column, so it is always visible while scrolling horizontally and/or vertically.

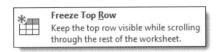

Before and after freezing top row

Freeze first row

1. Position the row to freeze at the top
2. Select **View** tab → **Freeze Panes** list → **Freeze Top Row**

Freeze first column

1. Position the column to freeze to the left
2. Select **View** tab → **Freeze Panes** list → **Freeze First Column**

Notice, the first 'visible' column or row is frozen

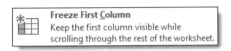

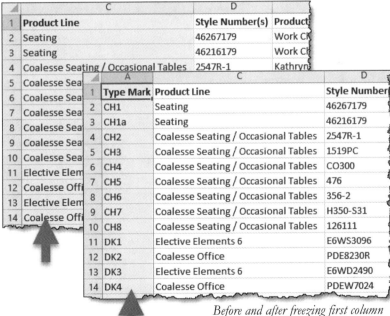

Before and after freezing first column

Unfreeze Panes

1. Select **View** → **Freeze Panes** → **Unfreeze Panes**

1.4.4 Change window views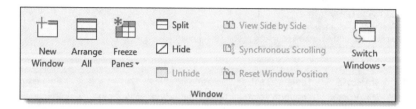

The window panel, on the View tab of the Ribbon, has multiple tools to control how the application and document windows are organized on the screen.

New Window

Opens a second window of the same document. This facilitates viewing/working in different parts of the document at the same time. This is especially helpful on a multi monitor system.

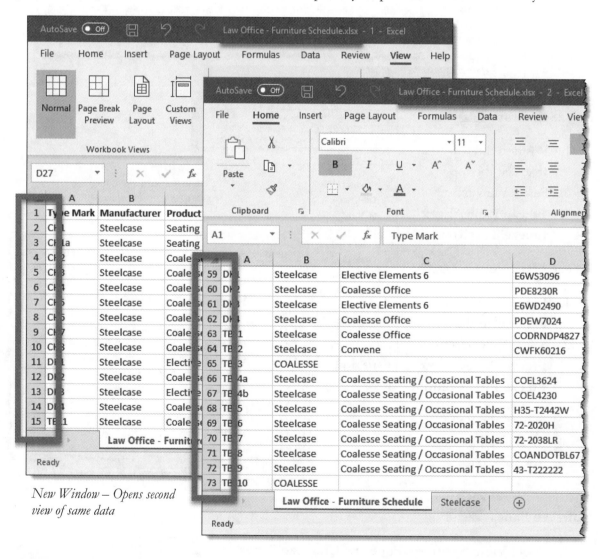

New Window – Opens second view of same data

Arrange All

The **Arrange All** feature changes the composition of all open Excel documents, aka workbooks, on the screen. Selecting this command opens the dialog shown to the right. The results of each of the four options are shown below.

These results can also be achieved manually by dragging the application title bar and adjusting its edges. Click the Maximize application icon in the upper right (on the application title bar) to return to full screen.

Arrange All dialog

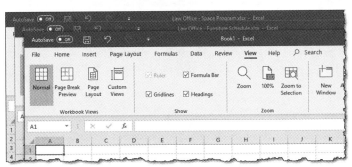

Arrange All – Cascade option

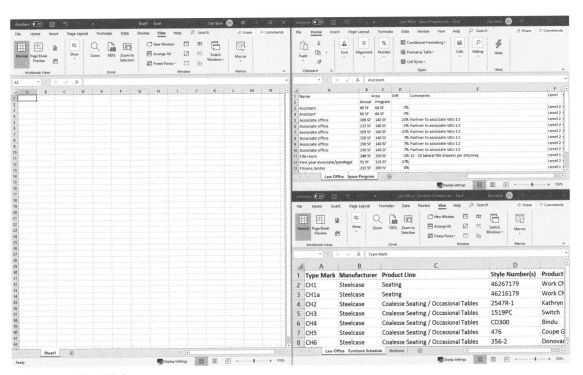

Arrange All – Tiled

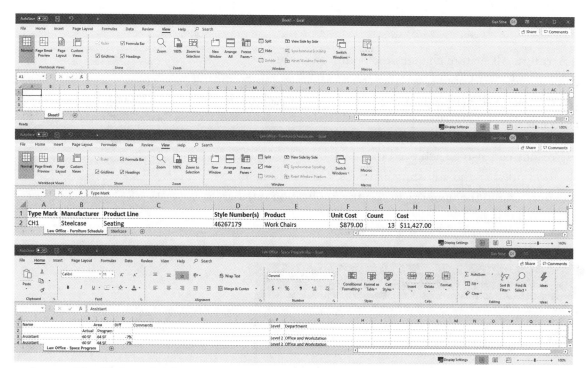

Arrange All – Horizontal option

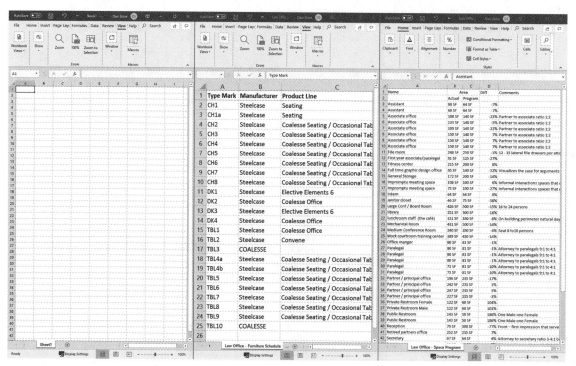

Arrange All – Vertical option

Split

To aid in comparing or presenting data, the sheet may be 'split' horizontally or vertically. It is possible to scroll on each side of the split. Select **Split** a second time to remove a split.

Split (horizontally)

1. Select row heading (split will occur above selection)
2. Select **View** tab ➔ **Split**

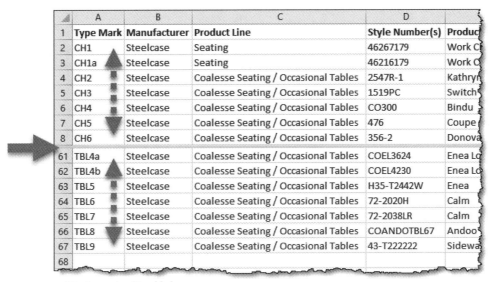

Horizontal split added

Split (vertically)

1. Select column heading (split will occur left of the selection)
2. Select **View** tab ➔ **Split**

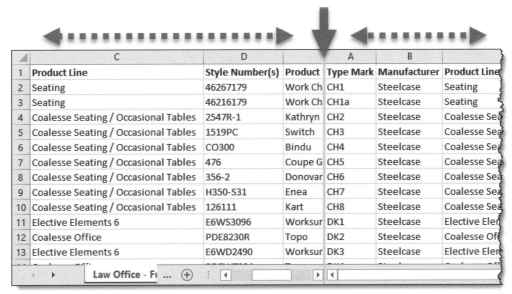

Vertical split added

1.4.5 Modify basic workbook properties

Each workbook, or file, has properties to help define its contents.

Access/edit workbook properties

1. Select the **File** tab
2. On the **Info** section, click the **Properties** title (see image)
3. Edit any of the five tabs in the Properties dialog

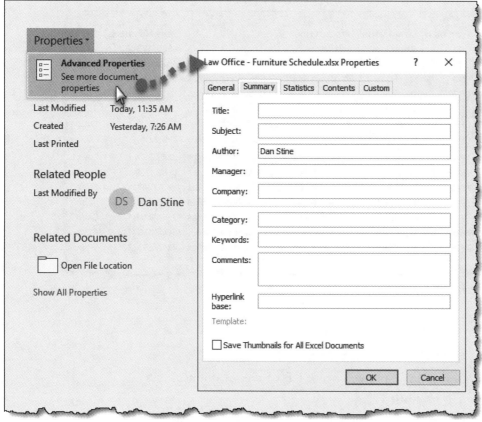

Properties dialog for current workbook

1.4.6 Display formulas

By default, a cell displays the result of a formula calculation. When the cell is selected, the formula can be seen in the Formula Bar. However, it is possible to see all formulas in the worksheet.

Display formulas

1. Select the **Formulas** tab
2. Click the **Show Formulas** button

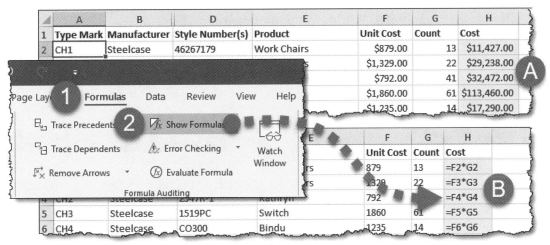

Before and after 'Show Formulas' toggle selected

> The column widths automatically adjust when toggling Show Formulas on and off. Double-click the space between the column headers to automatically space the column (on the left) to fit the data, i.e. make the row with the most information visible.

Display formula Results (default setting)

1. Select the **Formulas** tab
2. Click the highlighted **Show Formulas** button (to toggle it off)

1.5 Configure content for collaboration

This section covers steps used to share data with others and manage personal information.

1.5.1 Set a print area

To print less than all the information within a worksheet, define a print area.

Set a print area

1. **Click** in the **upper left** cell of the desired print area
2. **Drag** and release the mouse button in the **lower right** cell
3. Switch to the **Page Layout** tab
4. Expand the **Print Area** list
5. Select **Set Print Area**

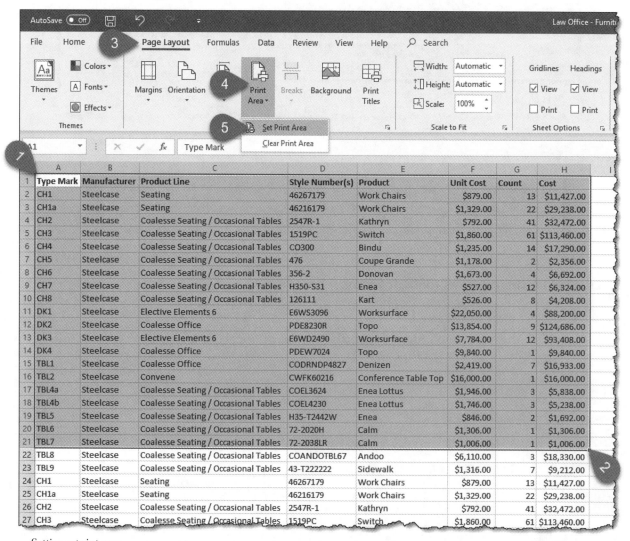

Setting a print area

The effects of setting a print area can be seen in the Page Layout and Page Break view types, as well as in a 'print preview' as shown below.

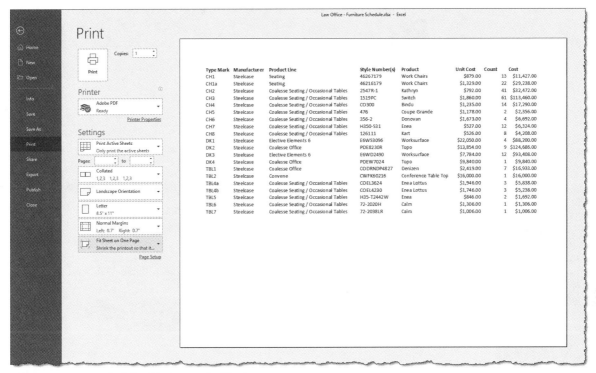

Print preview – only defined 'print area' will print

Once a print area is defined, it is possible to add additional, adjacent areas as shown in the image below. Each print area, within the same worksheet, will print on a separate page.

Add to print area

1. Select another range within a worksheet
2. Expand the **Print Area** list
3. Select **Add to Print Area**

To remove a defined print area, follow these steps.

Clear a print area

1. Switch to the **Page Layout** tab
2. Expand the **Print Area** list
3. Select **Clear Print Area**

Multiple print areas on same sheet, using Add to Print Area

1.5.2 Save workbooks in alternative file formats

Excel can save the workbook data in 29 different file formats. This is sometimes required to import data into other applications. The result is a new file with a different extension.

Save As alternate format 🎥

1. Select the **File** tab on the Ribbon
2. Click **Save As** on the left
3. Expand the **Save As Type** list
4. Select the desired **file type**

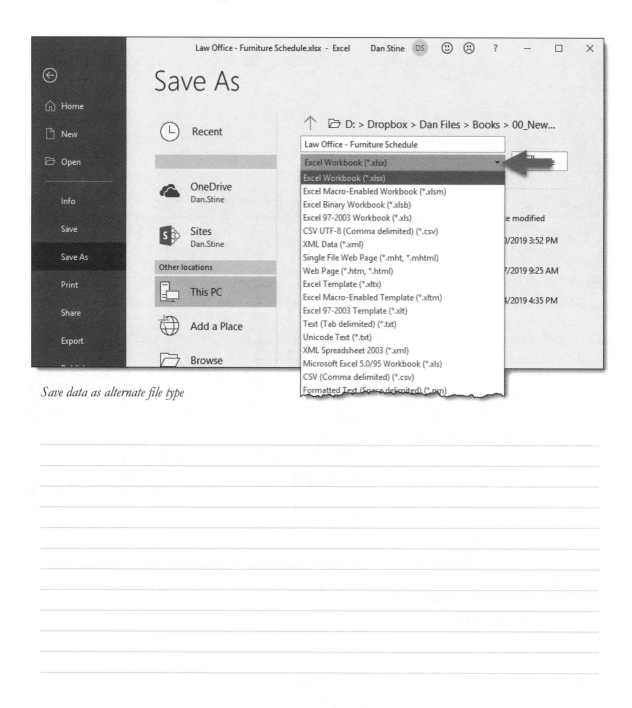

Save data as alternate file type

1.5.3 Configure Print Settings

There are many settings available when it comes to printing a worksheet. This section covers the highlights. This applies to hardcopies and PDFs. The print preview is updated automatically.

Access print settings

1. Select the **File** tab on the Ribbon
2. Click **Print** on the left
3. Adjust the settings per the information below
4. When finished
    a. Click **Print**, *or*
    b. Click **left arrow** to return to worksheet

Print settings outline:

1. **Print**: create hardcopy or PDF

2. **Copies**: number of copies

3. **Printer**: physical printer or PDF

4. **What to print**:
    a. Print Active Sheets
    b. Print Entire Workbook
    c. Print Selection

5. **Collate**: Yes or No

6. **Orientation**: Portrait or Landscape

7. **Paper size**: select from list based on selected printer

8. **Margins**: Normal, Wide, Narrow

9. **Scale**:
    a. No Scaling
    b. Fit Sheet on One Page
    c. Fit All Columns on One Page
    d. Fit All Rows on One Page

Click the **Print Properties** link for additional, printer specific, settings.

Click the Page Setup link to access the **Page Setup** dialog covered previously.

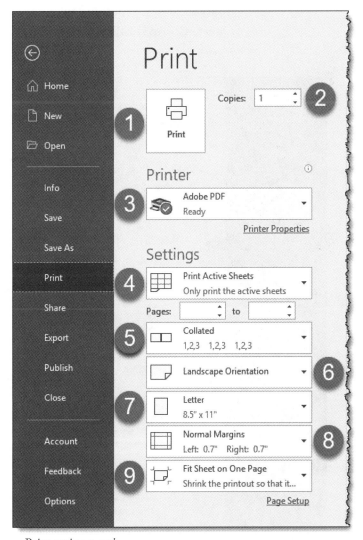

Print settings panel

1.5.4 Inspect workbook for issues

Before publishing a document, check to see what personal information it might contain, if it may be difficult for a person with a disability to read, or if it contains any features not supported by older versions of Excel.

Check for Issues 🎥

1. Select the **File** tab on the Ribbon
2. Click **Info** on the left
3. Expand the **Check for Issues** list
4. Select:
   a. **Inspect Document**: Check the workbook for hidden properties or personal information, with an option to 'Remove' for each section reported.
   b. **Check Accessibility**: Check the workbook for content that people with disabilities might find difficult to read.
   c. **Check Compatibility**: Check for features not supported by earlier versions of Excel.

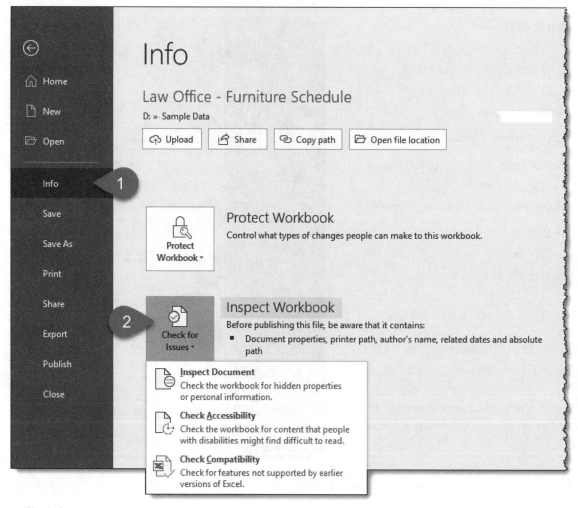

Check document for issues

1.0 Practice Tasks

Try the topics covered in this chapter to make sure you understand the concepts. These tasks are sequential and should be completed in the same Excel workbook unless noted otherwise. Saving the results is optional, unless assigned by an instructor.

Task 1.1:

✓ Open a new Excel workbook and **import** Furniture Schedule.txt.

Task 1.2

✓ Add a **hyperlink** to the first instance of Steelcase, pointed to: https://www.steelcase.com

Task 1.3

✓ Change the **page orientation** to Landscape in Page Setup.

Task 1.4:

✓ **Freeze** the first column.

Task 1.5:

✓ Set the **Print Area** to match the imported table.

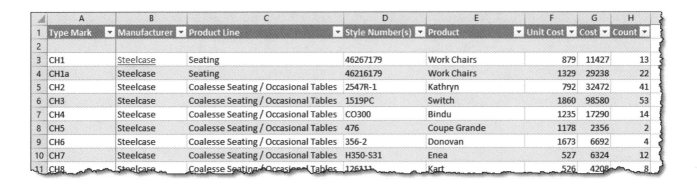

	A	B	C	D	E	F	G	H
1	Type Mark	Manufacturer	Product Line	Style Number(s)	Product	Unit Cost	Cost	Count
2								
3	CH1	Steelcase	Seating	46267179	Work Chairs	879	11427	13
4	CH1a	Steelcase	Seating	46216179	Work Chairs	1329	29238	22
5	CH2	Steelcase	Coalesse Seating / Occasional Tables	2547R-1	Kathryn	792	32472	41
6	CH3	Steelcase	Coalesse Seating / Occasional Tables	1519PC	Switch	1860	98580	53
7	CH4	Steelcase	Coalesse Seating / Occasional Tables	CO300	Bindu	1235	17290	14
8	CH5	Steelcase	Coalesse Seating / Occasional Tables	476	Coupe Grande	1178	2356	2
9	CH6	Steelcase	Coalesse Seating / Occasional Tables	356-2	Donovan	1673	6692	4
10	CH7	Steelcase	Coalesse Seating / Occasional Tables	H350-S31	Enea	527	6324	12
11	CH8	Steelcase	Coalesse Seating / Occasional Tables	126A11	Kart	526	4208	8

Notes:

2.0 Manage Data Cells and Ranges

Introduction

Use this chapter to review the certification outcomes around managing data and ranges. Use the provided files to practice the features and workflows to reinforce the concepts.

2.1. Manipulate data in worksheets

For the Excel certification exam, it is important to know how to copy/paste data, quickly fill cells with adjacent repeating or enumerated data and to insert/delete rows and columns. This section will cover these topics.

2.1.1 Paste data by using special paste options

Excel can Copy and Paste content from most Windows-based programs. But it also has a Paste Special feature set to insert attributes unique to spreadsheets, such as formulas, comments and more.

Basic Paste Special Steps

1. Select the cell(s) to copy
2. Press **Ctrl + C** (*or*, right-click and select Copy)
3. Select a cell(s) to paste to
4. **Home → Paste** (down arrow) → **Paste Special**
5. Select an option(s) and click **Ok**

> Some Paste Special commands are also found as icons in the Paste list shown below.

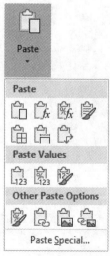

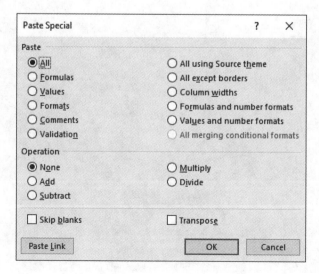

Paste Special Options

Paste Special actions defined

Paste	Action
All	All contents and attributes
Formulas	Just the formula; references are relative
Values	Current data or result of formula, but no formula
Formats	Cell formatting; e.g. bold, border, shading, etc.
Comments	Any comments associated with cell, via Review → New Comment
Validation	Data validation rules; via Data → Data Validation
All using Source theme	Cell theme applied to incoming data
All except borders	Same as "All" except border formatting is omitted
Column widths	Column width of 'copied' cell applied to select cell/column
Formulas and number formats	Formula and number formatting from Format Cells/Number tab
Values and number formats	Formula result and number formatting

Operation	Action
None	Normal paste per options above
Add	Paste '2' into cell containing '4' results in '6'
Subtract	Paste '2' into cell containing '4' results in '2'
Multiply	Paste '2' into cell containing '4' results in '8'
Divide	Paste '2' into cell containing '4' results in '2'
Validation	Validates data being pasted
Skip blanks	Blank cells in a selected range are ignored, possibly leaving old data in paste-area selection
Transpose	Paste columns into rows or rows into columns
Paste Link	Creates a link between copied and pasted data

2.1.2 Fill cells by using Auto Fill

The Auto Fill feature quickly populates adjacent cells with the selected data or pattern of data. If the selected cell contains a formula, the formula will be copied and any references to other cells will be relative; thus, the formula is not copied verbatim (unless using 'absolute' references).

Auto Fill the Same Value

1. Select the cell to copy
2. Click and drag the lower right grip down (up or over)

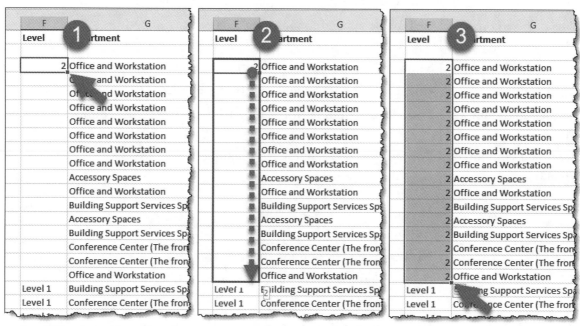

Auto Fill a single value or formula

Auto Fill the Value Pattern

1. Select the cells which define the pattern to copy
2. Click and drag the lower right grip down (up or over)

Pattern Examples:

1,2 *result* 1,2,3,4,5,6…

2,4,6 *result* 2,4,6,8,10…

Auto Fill a pattern or series

If the selected cell contains text, by default Excel will increment the number in the select cell. See the image below for some examples of cells with text and numbers.

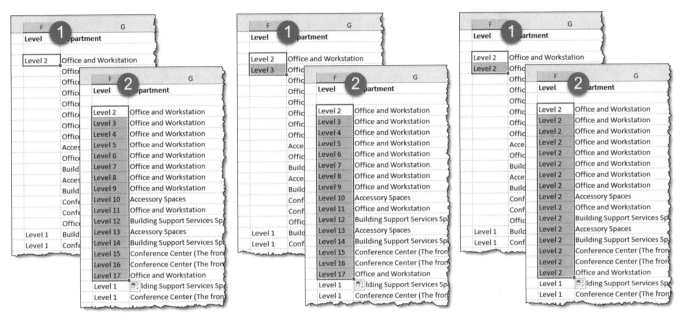

Auto Fill using a combination of text and numbers

2.1.3 Insert and delete multiple columns or rows

Follow these steps to insert multiple blank rows or columns into a worksheet.

Insert multiple columns

1. Select the number of columns to insert by dragging on headers, i.e. the column letter(s) or the row number(s)
2. **Right-click** a selected column header
3. Select **Insert**

> The contents of the selected rows/columns are <u>not</u> copied.

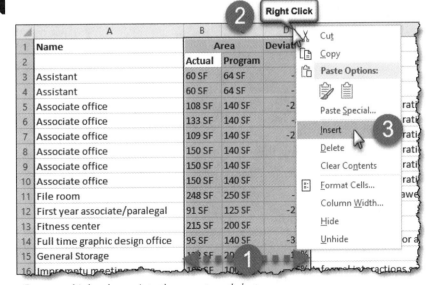

Insert multiple columns into the current worksheet

The new columns are inserted to the left of the selected columns. The steps are similar for rows, with new blank rows being added directly above the current selection.

2.1.4 Insert and delete cells

It is possible to insert or delete individual cells, but caution should be used as this can cause data to shift out of alignment with adjacent data.

Insert A Single Cell

1. **Right-click** on a cell
2. Select **Insert** from the menu
3. **Insert dialog:** select an option carefully
4. A new cell has been inserted; **review the results**

Notice in this example, the 'Deviation' column appears to not be 'calculated values' so a significant error was created by shifting the data downward to make room for a new cell.

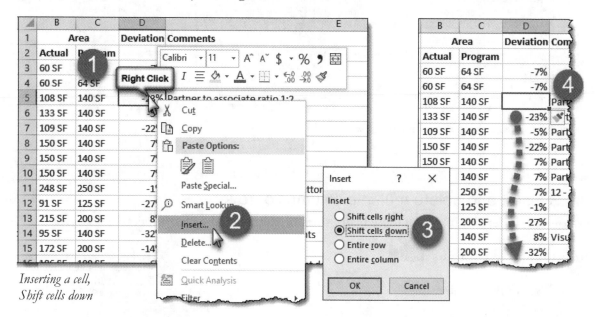

Inserting a cell,
Shift cells down

The image to the right shows the results of shifting the data to the right.

Delete A Cell(s):
The steps are the same to delete a cell, except the cells are shifted in the opposite direction.

Inserting a cell, Shift cells right

2.2. Format Cells and Ranges

The Excel certification exam requires an understanding about formatting cells and ranges.

2.2.1 Merge and unmerge cells

To better organize or visualize data, cells can be merged together to form a larger cell.

Merge Cells

1. **Identify** cells to merge
2. **Select** cells to merge; e.g. B1:C1
3. **Home** (tab) → **Merge & Center** (arrow) → **Merge & Center**
4. If the selected cells contain data: OK to acknowledge data may be lost
5. Cells are now merged

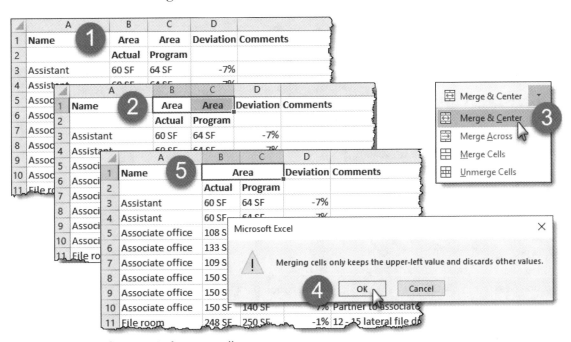

Steps required to merge cells

The image to the right shows additional cells merged to create an attractive header for the data presented.

Unmerge Cells

1. **Select** cell to unmerge
2. **Home** → **Merge & Center** (arrow) → **Unmerge Cells**

	A	B	C	D
1		Area		
2	Name	Actual	Program	Deviation
3	Assistant	60 SF	64 SF	-7%
4	Assistant	60 SF	64 SF	-7%
5	Associate office	108 SF	140 SF	-23% Pa
6	Associate office	133 SF	140 SF	-5% Pa

Example of multiple groups of cells merged

2.2.2 Modify cell alignment, orientation, and indentation

The data in a selected cell or range can be aligned vertically or horizontally, as well as indented and rotated. The images below describe the options using tools on the Home tab for selected data.

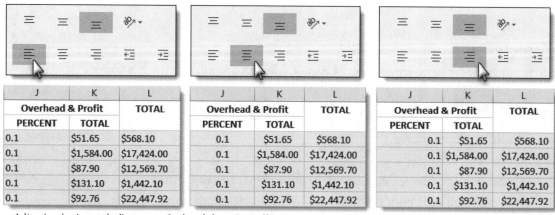

Adjusting horizontal alignment of selected data (i.e. cells)

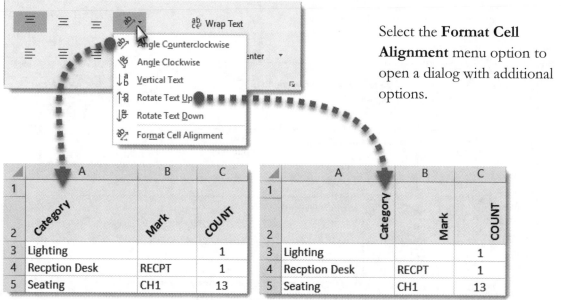

Select the **Format Cell Alignment** menu option to open a dialog with additional options.

Adjusting orientation of cell data

Each time the **Indent** icon is clicked, the selected data moves right. Conversely, clicking the **Unindent** moves the selected data left.

Data may also be positioned vertically using the three icons on the top row.

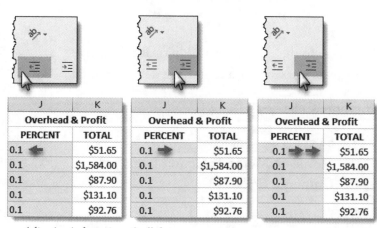

Adjusting indentation of cell data

2.2.3 Format cells by using Format Painter

Quickly copy the formatting properties, not the data, from one cell/range to another.

Copy Formatting using Format Painter

1. **Select** cell, or range, to copy formatting settings from
2. Click **Home → Format Painter** from the Ribbon
3. **Select** cell, or range, to apply formatting properties to

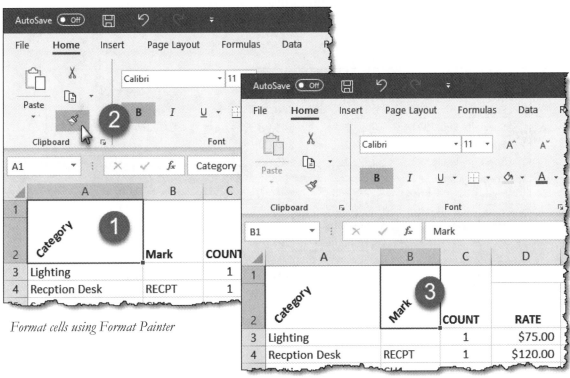

Format cells using Format Painter

Apply Same Formatting using Format Painter to Multiple Locations

1. **Select** cell, or range, to copy formatting settings from
2. Double-Click **Format Painter** from the Home tab on the Ribbon
3. Select cells/ranges to apply formatting
4. Press **Esc** to finish

2.2.4 Wrap text within cells

If the data in a cell does not fit within the width of the column it is obscured. Text is clipped at the right edge of the column and numbers show as "#####". To resolve this issue, either the column needs to be made wider or text wrapping needs to be enabled for specific locations. Note that text wrapping causes individual rows to become taller.

Wrap text in selected cell or range 🎥

1. **Select** cell, column heading or range to adjust
2. Click **Home → Wrap Text** from the Ribbon
3. Review results

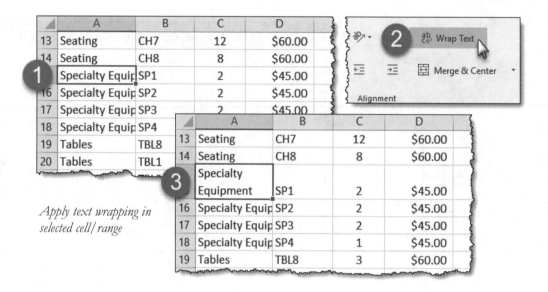

Apply text wrapping in selected cell/range

2.2.5 Apply number formats

Cells can be formatted to represent specific types of data, such as currency, scientific, time, date, etc. Doing so helps to validate data entry and automates the formatting.

Apply Number Formats

1. **Select** a cell, range, column or row heading
2. Select from the **Home → Number list**
3. Click **More Number Formats** for more options

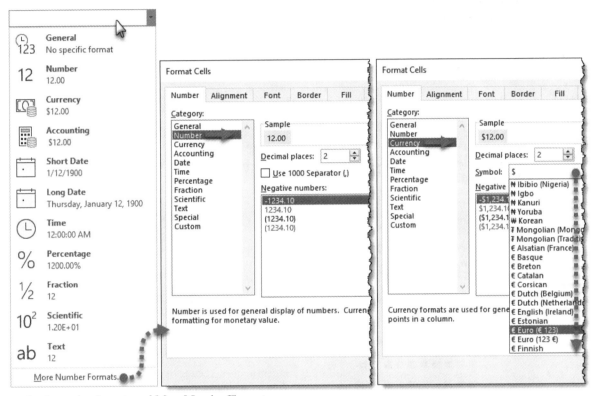

Apply number formats and More Number Formats

When a single cell is selected and contains a number, e.g. 12, the **Number** list, on the Home tab, will display the various ways in which this number may be represented as shown in the example above.

In the **Format Cells** dialog, on the Number tab, one will find additional options such as number of decimal places, add a separator every three digits, which currency symbol to use based on a large list of world currency options, and more.

General is the default number setting, allowing for any data, including letters.

2.2.6 Apply cell formats from the Format Cells dialog box

When it comes to formatting cells in Excel, the most options can be found in the Format Cells dialog. Each tab in this dialog shows sample results based on the pending changes.

Use the Format Cells dialog

1. **Select** a cell, range, column or row heading
2. Right click and click **Format Cells...** from the menu
3. Make adjustments on one or more of the **tabs**
4. Click **OK** to apply the changes.

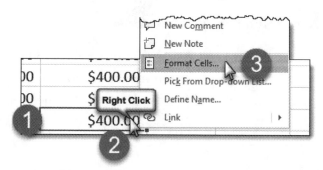

Access Format Cells dialog via right-click

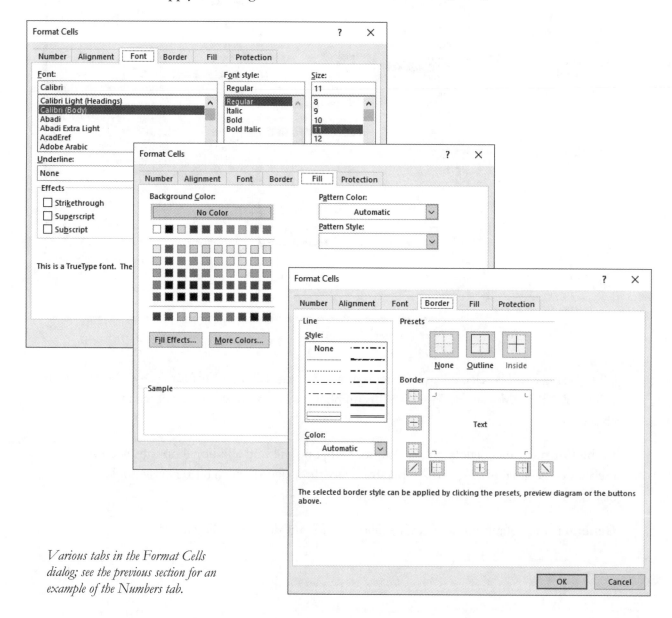

Various tabs in the Format Cells dialog; see the previous section for an example of the Numbers tab.

2.2.7 Apply cell styles

A Cell Style is a collection of format settings. Using them makes it easier to manage multiple formatting strategies throughout a workbook. Any changes to a Cell Style instantly updates all locations.

Applying cell styles

1. **Select** a cell, range, column or row heading
2. Click **Home → Cell Styles**
3. **Select** an option

See five examples to the right. In the Styles list (see image below), **Right-click → Modify** to edit a style. Select **New Cell Styles…** to open the Style dialog and create a new cell style.

The **Normal** cell style will remove all formatting, including formatting applied prior to a cell style being used.

Sample Cell Styles applied

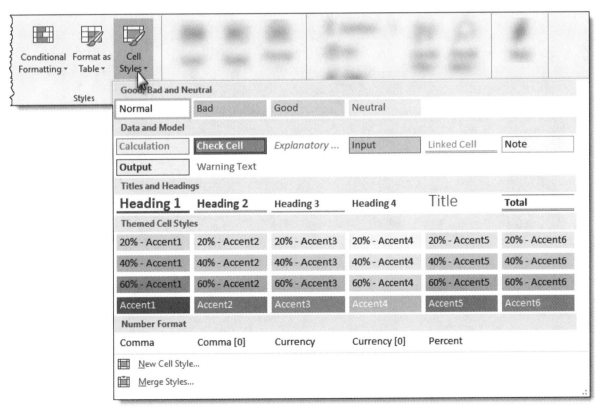

Cell Styles list from Home tab on the Ribbon

2.2.8 Clear cell formatting

It is easy to remove all cell formatting and reduce the selection to just the original data, or formula, entered.

Clear cell formatting of selected area

1. **Select** a cell, range, column or row heading
2. Click **Home → Clear → Clear Formats**

Take a moment to notice the other "Clear" options listed in the image to the right.

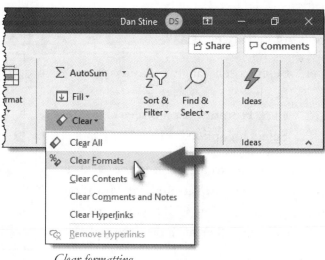

Clear formatting

2.3. Define and reference named ranges

This section covers how selected areas (i.e. a range) and tables are named.

2.3.1 Define a named range

Naming a range makes it easy to reselect, or reference, a specific data set. In the following example all the rooms on Level 1 are selected and named "Level_1".

Define a named range 🎥

1. **Select** a range; click in the upper left cell
2. **Select** a range; drag to the lower right cell
3. Type a new name in the **Name Box**
    a. Spaces & certain symbols are not allowed

Once a range is named, it can be selected via the Name Box list as shown in the image to the right.

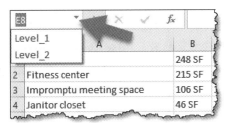

Select a named range

	A	B	C	D	E	F	G
1	F...om	248 SF	250 SF	-1%	12 - 15 lateral file drawers per	Level 1	Accessory Spaces
2	...center	215 SF	200 SF	8%		Level 1	Building Support Services Spaces
3	Impromptu meeting space	106 SF	100 SF	6%	Informal interactions spaces th	Level 1	Conference Center (The front entry)
4	Janitor closet	46 SF	75 SF	-38%		Level 1	Building Support Services Spaces
5	Large Conf / Board Room	426 SF	500 SF	-15%	16 to 24 persons	Level 1	Conference Center (The front entry)
6	library	251 SF	300 SF	-16%		Level 1	Accessory Spaces
7	lunchroom staff (the café)	321 SF	350 SF	-8%	On building perimeter natural	Level 1	Accessory Spaces
8	Mechanical Room	431 SF	500 SF	-14%		Level 1	Building Support Services Spaces
9	Medium Conference Room	240 SF	250 SF	-4%	Seat 8 to10 persons	Level 1	Conference Center (The front entry)
10	Mock courtroom training center	389 SF	450 SF	-14%		Level 1	Accessory Spaces
11	Public Restroom	143 SF	50 SF	186%	One Male one Female	Level 1	Conference Center (The front entry)
12	Public Restroom	143 SF	50 SF	186%	One Male one Female	Level 1	Conference Center (The front entry)
13	Reception	70 SF	300 SF	-77%	Front – first impression that se	Level 1	Conference Center (The front entry)
14	Small Conference Room	123 SF	100 SF	23%	Office size that could change u	Level 1	Conference Center (The front entry)
15	Small Conference Room	102 SF	100 SF	2%	Office size that could change u	Level 1	Conference Center (The front entry)
16	Vestibule	297 SF	300 SF	-1%		Level 1	Conference Center (The front entry)
17	Visitor office	118 SF	125 SF	-6%		Level 1	Office and Workstation
18	Waiting / lobby	304 SF	300 SF	1%		Level 1	Conference Center (The front
19	Workroom	342 SF	300 SF	14%		Level 1	Building Support Services Space
20	Assistant	60 SF	64 SF	-7%		Level 2	Office and Workstation
21	Assistant	60 SF	64 SF	-7%		Level 2	Office and Workstation
22	Associate office	108 SF	140 SF	-23%	Partner to associate ratio 1:2	Level 2	Office and Workstation
23	Associate office	133 SF	140 SF	-5%	Partner to associate ratio 1:2	Level 2	Office and Workstation

Name Box: Level_1 | Building Support Services Spaces

Define a named range

2.3.2 Name a table

Tables have a generic name by default, but the name may be changed for clarity.

Name a table 🎥

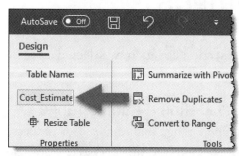

1. **Select** anywhere within a table
2. **Click** the Table Design tab on the Ribbon
3. Edit the **Table Name** box
   a. Spaces & certain symbols are not allowed

Name a table from the Design tab

To navigate to, and select, the named table simply select it from the Name Box as shown in the image to the right.

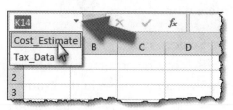

Navigate to named table

> Named ranges or tables can be used in formulas. For example, if a range or table is named **QuarterlyEarnings**, then use the following formula to gather a total: **=SUM(QuarterlyEarnings)**

2.4. Summarize Data Visually

A powerful feature of Excel is its ability to make data more understandable through formatting.

2.4.1 Insert sparklines

A sparkline is a single cell-sized graph depicting relative changes in adjacent data. In the example to the right, rows convey associate sales per quarter, while the columns represent overall company sales.

Insert a Sparkline

1. **Select** a cell adjacent to data
2. Click **Insert → Line** (Sparkline)
3. Create Sparklines dialog: **Select range** in worksheet
4. Click **OK**
5. Review the results

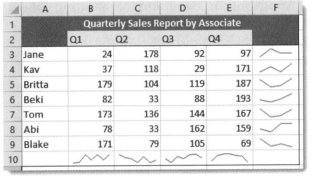

Sparkline example for both rows and columns

Edit options are available by right-clicking on a sparkline, via the sparkline fly-out menu.

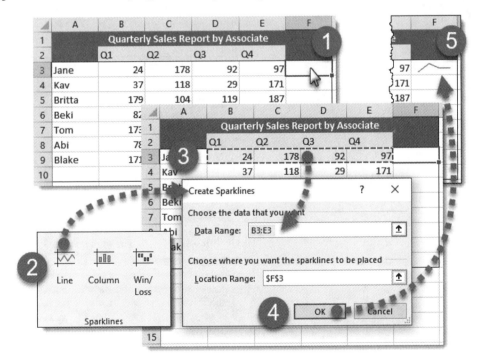

Inserting a Sparkline

2.4.2 Apply built-in conditional formatting

Excel can change the way data appears based on certain conditions; this is called conditional formatting. Several examples are shown to the right. Conditional formatting is dynamic and will automatically change if the data is modified.

Apply built-in conditional formatting

1. **Select** cell, range, column or row
2. Click **Home → Conditional Formatting**
3. Select an option

Some options have intermediate steps. For example, selecting **Greater Than…** displays a dialog to answer the question "Greater than what?"

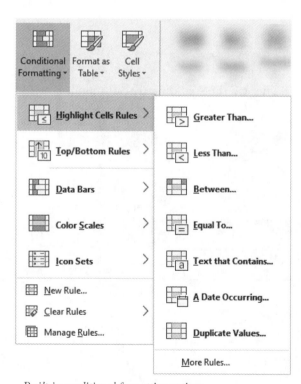

Built-in conditional formatting options

Bottom 10 %…

	Q1	Q2	Q3	Q4	
Jane	24	178	92	97	
Kav	37	118	29	171	
Britta	179	104	119	187	
Beki	82	33	88	193	
Tom	173	136	144	167	
Abi	78	33	162	159	
Blake	171	79	105	69	

Sales Report by Associate

Top 10 %…

	Q1	Q2	Q3	Q4	
Jane	24	178	92	97	
Kav	37	118	29	171	
Britta	179	104	119	187	
Beki	82	33	88	193	
Tom	173	136	144	167	
Abi	78	33	162	159	
Blake	171	79	105	69	

Sales Report by Associate

Greater Than…

	Q1	Q2	Q3	Q4	
Jane	24	178	92	97	
Kav	37	118	29	171	
Britta	179	104	119	187	
Beki	82	33	88	193	
Tom	173	136	144	167	
Abi	78	33	162	159	
Blake	171	79	105	69	

Sales Report by Associate

Data Bars

	Q1	Q2	Q3	Q4	
Jane	24	178	92	97	
Kav	37	118	29	171	
Britta	179	104	119	187	
Beki	82	33	88	193	
Tom	173	136	144	167	
Abi	78	33	162	159	
Blake	171	79	105	69	

Sales Report by Associate

Icon Sets

	Q1	Q2	Q3	Q4	
Jane	24	178	92	97	
Kav	37	118	29	171	
Britta	179	104	119	187	
Beki	82	33	88	193	
Tom	173	136	144	167	
Abi	78	33	162	159	
Blake	171	79	105	69	

Sales Report by Associate

Conditional formatting examples

2.4.3 Remove conditional formatting

At some point, conditional formatting may not be needed. Here is how it is cleared.

Remove conditional formatting

1. **Select** cell, range, column or row with conditional formatting applied
2. Click **Home → Conditional Formatting → Clear Rules**
3. Select an option:
   a. **Clear Rules from Selected Cells**
   b. Clear Rules from Entire Sheet

The conditional formatting is now removed from the cell/worksheet.

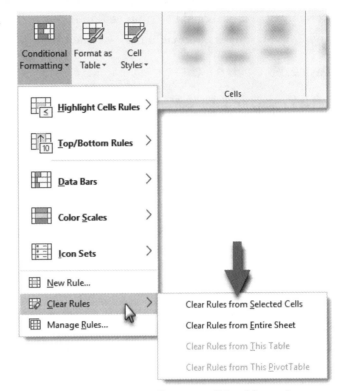

Remove conditional formatting

2.0 Practice Tasks

Try the topics covered in this chapter to make sure you understand the concepts. These tasks are sequential and should be completed in the same Excel workbook unless noted otherwise. Saving the results is optional, unless assigned by an instructor.

Task 2.1:

✓ Open Monthly Sales Report by Associate.xlsx, and **insert** two rows at once, above Beki.

Task 2.2

✓ **Rotate** the "month headers" text 45 degrees.

Task 2.3

✓ Define a **named range**: select the range A1:N11 and name it FullTable.

Task 2.4:

✓ Insert a **Spark Line** at N3 that represents the first row, Jane.

Task 2.5:

✓ Use the **AutoFill** feature to copy the Spark Line down for every row.

	A	B	C	D	E	F	G	H	I	J	K	L	M	N
1						Monthly Sales Report by Associate								
2		Jan	Feb	Mar	Apr	May	Jun	Jul	Aug	Sep	Oct	Nov	Dec	
3	Jane	186	108	92	122	190	71	21	37	24	178	92	97	
4	Kav	15	16	198	44	25	68	43	119	37	118	29	171	
5	Britta	166	185	89	170	131	70	50	149	179	104	119	187	
6														
7														
8	Beki	21	113	83	17	130	26	167	102	82	33	88	193	
9	Tom	70	160	125	84	191	97	52	45	173	136	144	167	
10	Abi	61	99	70	162	28	163	101	103	78	33	162	159	
11	Blake	105	55	163	12	117	83	163	120	171	79	105	69	

3.0 Manage Tables and Table Data

Introduction

This chapter covers the required outcomes related to managing tables and table data.

3.1.   Create and Format Tables

In Excel, Tables are a subset of data within a worksheet which make managing and analyzing related data easier.

3.1.1 Create Excel tables from cell ranges

This objective requires one to know how to create a table from a range.

Create table from cell ranges

1. **Select** the upper left cell of area to be defined as a table
2. Click **Format as Table** → *any option*
3. Dialog: **Select range** directly in workbook, or type range; click **OK**
4. **Review** results

In the example below, row one contains header information (it is not part of the data) so **My table has headers** was left checked in step #3. The headers now contain special filter arrows.

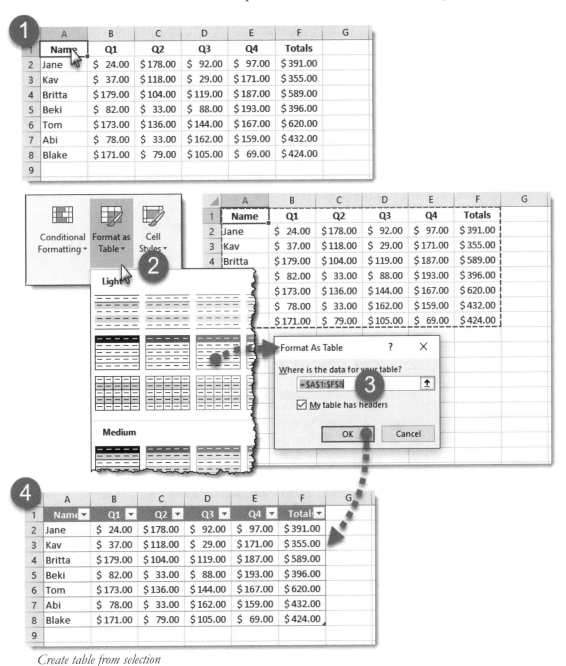

Create table from selection

3.1.2 Apply table styles

The steps just given to create a table also applied a table style. This topic provides instruction on changing the current table style.

Apply or change a table style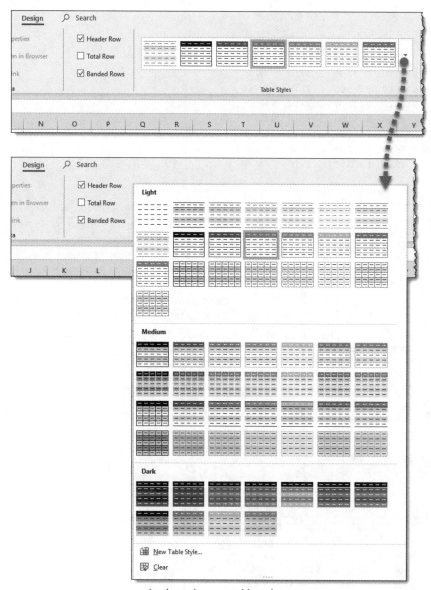

1. **Select** anywhere within a table
2. Click **Table Design → Table Styles** list → *any option*

> Hover cursor over an option to see the **style name** in a tooltip; the exam may specify a specific style to apply (notice the three groupings as well: Light, Medium and Dark).

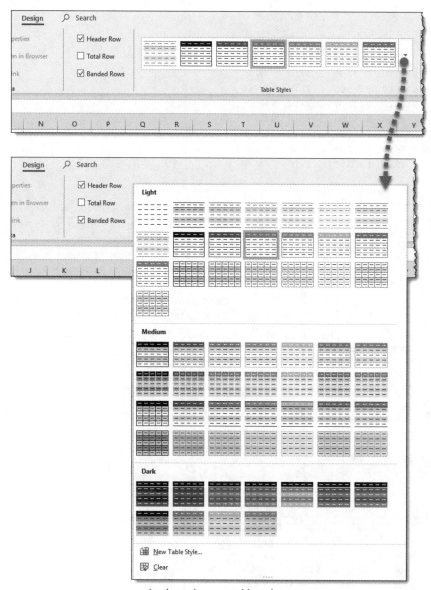

Apply or change a table style

3.1.3 Convert tables to cell ranges

Sometimes it is necessary to change a table back into normal cells, which removes the table.

Convert table to cell range 📹

1. **Select** anywhere within a table
2. Click **Table Design → Convert to Range**
3. Dialog: Click **OK** to convert entire table
4. **Review** results

The table no longer exists within the workbook. The data and formatting have been preserved.

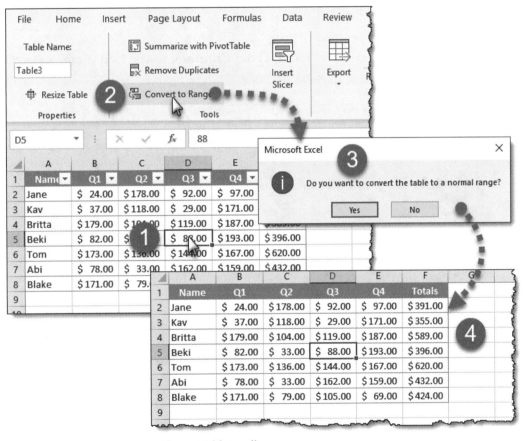

Convert table to cell range

3.2.   Modify tables

Review the required objectives in this section on modifying tables in a workbook.

3.2.1 Add or remove table rows and columns

Review how to insert or delete a row or column in an existing table.

Add table row or column

1. **Select** a cell(s) within a table; *row will be inserted above*
2. **Right-click** and select **Insert → Table Rows Above**
3. **Review** results

Steps are similar for columns; in step #2 select **Table Columns to the Left**. In the example below, notice data outside the table (e.g. the number 23) did not shift.

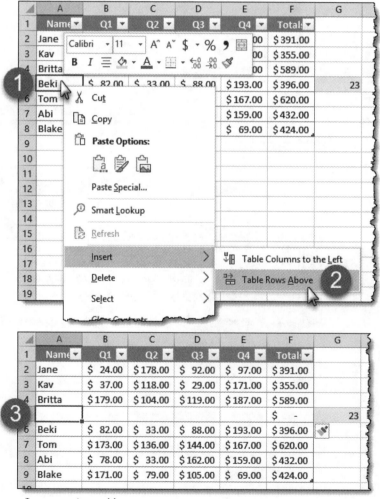

Insert row into table

Delete table row or column

1. **Select** a cell(s) within a table
2. **Right-click** and select **Delete → Table Rows**
3. **Review** results

Steps are similar for removing columns; in step #2 select **Table Columns**. In the example below, notice data outside the table (e.g. the number 23) did not shift.

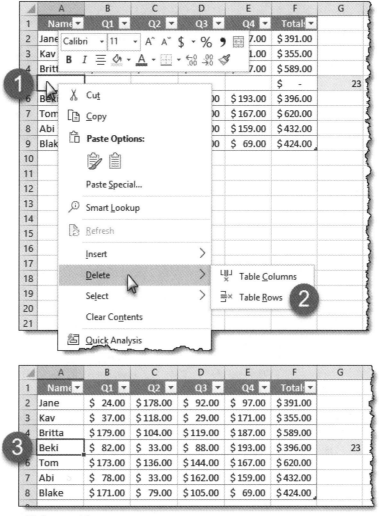

Delete a table row

3.2.2 Configure table style options

Tables have several built-in table style options to control how the table appears. These items may be toggled on and off without losing any data.

Toggle table style options

1. **Select** anywhere within a table
2. Select the **Table Design** tab on the Ribbon; *a table must be selected to see this tab*
3. **Check** or **uncheck** the options in the **Table Style Options** panel

☑ Header Row	☐ First Column	☑ Filter Button
☐ Total Row	☐ Last Column	
☑ Banded Rows	☐ Banded Columns	

Table Style Options

Table style options on Design tab (table must be selected first)

The image below compares the same table with no **Table Style Options** applied (top) and with all options selected (bottom). Many variations are possible by selecting different combinations of options.

	A	B	C	D	E	F	
1							
2	Jane	$ 24.00	$178.00	$ 92.00	$ 97.00	$ 391.00	
3	Kav	$ 37.00	$118.00	$ 29.00	$171.00	$ 355.00	
4	Britta	$179.00	$104.00	$119.00	$187.00	$ 589.00	
5	Beki	$ 82.00	$ 33.00	$ 88.00	$193.00	$ 396.00	
6	Tom	$173.00	$136.00	$144.00	$167.00	$ 620.00	
7	Abi	$ 78.00	$ 33.00	$162.00	$159.00	$ 432.00	
8	Blake	$171.00	$ 79.00	$105.00	$ 69.00	$ 424.00	
9							

	A	B	C	D	E	F	
1	Name	Q1	Q2	Q3	Q4	Totals	
2	Jane	$ 24.00	$178.00	$ 92.00	$ 97.00	$ 391.00	
3	Kav	$ 37.00	$118.00	$ 29.00	$171.00	$ 355.00	
4	Britta	$179.00	$104.00	$119.00	$187.00	$ 589.00	
5	Beki	$ 82.00	$ 33.00	$ 88.00	$193.00	$ 396.00	
6	Tom	$173.00	$136.00	$144.00	$167.00	$ 620.00	
7	Abi	$ 78.00	$ 33.00	$162.00	$159.00	$ 432.00	
8	Blake	$171.00	$ 79.00	$105.00	$ 69.00	$ 424.00	
9	Total					$ 3,207.00	

Compare table style options; no options v. all options applied

3.2.3 Insert and configure total rows

Inserting row totals, in a table, is part of the table style options that deserves special attention.

Insert table total row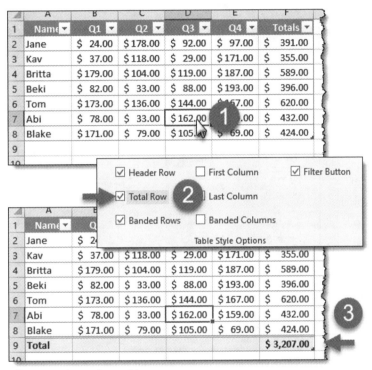

1. **Select** anywhere within a table
2. Select the **Table Design** tab on the Ribbon; *a table must be selected to see this tab*
3. Check the **Total Row** option on the Table Style Options panel

A new row is added to the bottom of the table as shown in the example below, listing a total.

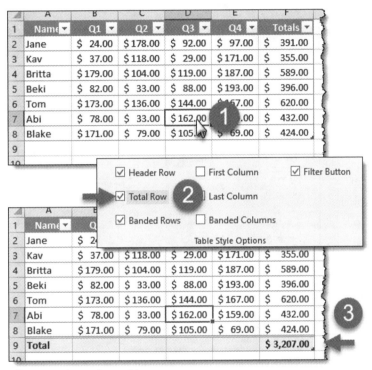

Insert total rows in a table

Optional step: Once a table Total Row is added, click at the bottom of each column (#4) to apply a function, such as sum, max, min, etc. (#5) to the column as shown in the image below.

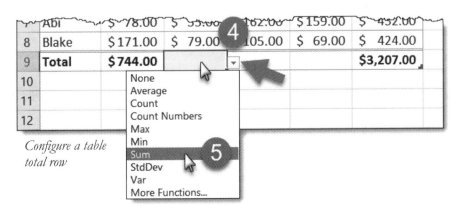

Configure a table total row

3.3.   Filter and sort table data

Tables can be quickly sorted and filtered to look at relevant data.

3.3.1 Filter records

When a table has the Header Row and Filter Button (turned on via Table Style Options) filtering data is very simple.

Filter records

1. **Click a filter arrow in the header row.**
2. Select a filter option(s).
3. Click **OK**.

Unlike sorting a regular column, the entire table is automatically sorted to maintain data integrity.

The image below compares the same table sorted/filtered three different ways.

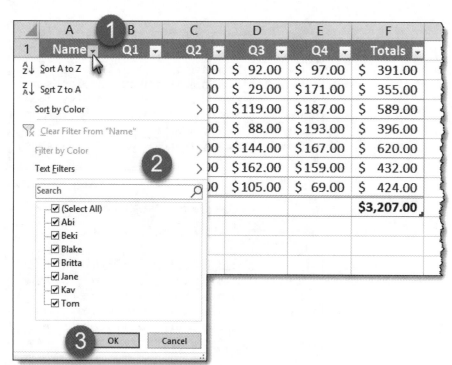

Click column arrow to view filter options

	A		
1	Name	Q1	C
2	Abi	$ 78.00	$ 3
3	Beki	$ 82.00	$ 3
4	Blake	$171.00	$ 7
5	Britta	$179.00	$10
6	Jane	$ 24.00	$17
7	Kav	$ 37.00	$11
8	Tom	$173.00	$13
9	**Total**		

	A	
1	Name	
3	Beki	
4	Blake	
5	Britta	
9	**Total**	
10		
11		
12		
13		

	A		
1	Name	Q1	C
2	Britta	$179.00	$10
3	Tom	$173.00	$13
4	Blake	$171.00	$ 7
5	Beki	$ 82.00	$ 3
6	Abi	$ 78.00	$ 3
7	Kav	$ 37.00	$11
8	Jane	$ 24.00	$17
9	**Total**		

Compare various filter results

3.3.2 Sort data by multiple columns

Tables may be sorted by more than one column in Excel.

In the example below, there are multiple categories with the same name, and they are further defined by the Mark category. Thus, we will sort by the Categories column and then the Mark column; e.g. seating will self-sort for Mark.

Sort data by multiple columns

1. **Click** anywhere within a table
2. Select **Home → Sort & Filter → Custom Sort…**
3. In the Sort dialog:
   a. Click the **Add Level** button
   b. Set the first 'Sort by' row to **Category**
   c. Set the second 'Sort by' row to **Mark**
   d. Click **OK**
4. Review the results

	A	B	L	
1	Category	Mark	Total	
2	Lighting	LT1	$568.10	
3	Seating	CH1	$12,569.70	
4	Specialty Equipment	SP1	$524.70	
5	Specialty Equipment	SP2	$999.90	
6	Specialty Equipment	SP3	$1,316.70	
7	Specialty Equipment	SP4	$480.15	
8	Seating	CH1a	$1,442.10	
9	Seating	CH1b	$22,447.92	
10	Seating	CH4	$13,203.96	
11	Seating	CH5	$2,646.60	
12	Seating	CH6	$6,547.20	
13	Seating	CH7	$10,890.00	
14	Seating	CH8	$8,162.88	
15	Recption Desk	RECPT	$17,424.00	
16	Tables	TBL8	$1,623.60	
17	Tables	TBL1	$6,560.40	
18	Tables	TBL2a	$557.70	
19	Tables	TBL2b	$4,474.80	
20	Seating	CH2	$25,030.50	
21	Seating	CH3	$73,756.32	
22	Tables	TBL3	$1,624.26	
23	Tables	TBL4b	$3,049.20	

	A	B	L	
1	Category	Mark	Total	
2	Lighting	LT1	$568.10	
3	Recption Desk	RECPT	$17,424.00	
4	Seating	CH1	$12,569.70	
5	Seating	CH1a	$1,442.10	
6	Seating	CH1b	$22,447.92	
7	Seating	CH2	$25,030.50	
8	Seating	CH3	$73,756.32	
9	Seating	CH4	$13,203.96	
10	Seating	CH5	$2,646.60	
11	Seating	CH6	$6,547.20	
12	Seating	CH7	$10,890.00	
13	Seating	CH8	$8,162.88	
14	Specialty Equipment	SP1	$524.70	
15	Specialty Equipment	SP2	$999.90	
16	Specialty Equipment	SP3	$1,316.70	
17	Specialty Equipment	SP4	$480.15	
18	Tables	TBL1	$6,560.40	
19	Tables	TBL2a	$557.70	
20	Tables	TBL2b	$4,474.80	
21	Tables	TBL3	$1,624.26	
22	Tables	TBL4b	$3,049.20	
23	Tables	TBL8	$1,623.60	

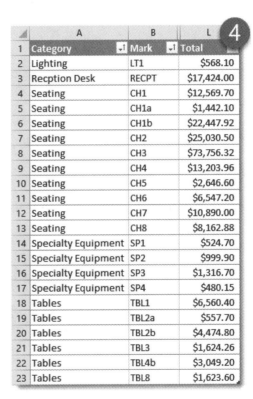

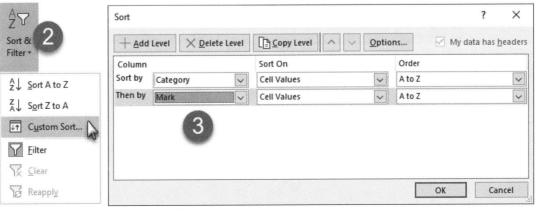

Sort data by multiple columns in table

3.0 Practice Tasks

Try the topics covered in this chapter to make sure you understand the concepts. These tasks are sequential and should be completed in the same Excel workbook unless noted otherwise. Saving the results is optional, unless assigned by an instructor.

Task 3.1:

✓ Open Quarterly Sales Report by Associate.xlsx, and then **create a table** based on the following range: A2:F9. Use the following style: Orange, Table Style Light 10.

Task 3.2

✓ **Insert** a new table row above Abi.

Task 3.3

✓ Add a **Total Row** using Table Style Options.

Task 3.4:

✓ Toggle on the **First Column** setting via Table Style Options.

Task 3.5:

✓ **Sort** the table by first name, alphabetically.

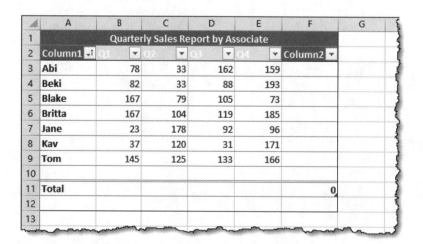

	A	B	C	D	E	F	G
1	Quarterly Sales Report by Associate						
2	Column1	Q1	Q2	Q3	Q4	Column2	
3	Abi	78	33	162	159		
4	Beki	82	33	88	193		
5	Blake	167	79	105	73		
6	Britta	167	104	119	185		
7	Jane	23	178	92	96		
8	Kav	37	120	31	171		
9	Tom	145	125	133	166		
10							
11	Total					0	
12							
13							

4.0 Perform Operations by using Formulas and Functions

Introduction

In this chapter the steps required to perform operations by using formulas and functions are reviewed. What's the difference between a formula and function? With a cell selected, in the Formula Bar, a simple <u>formula</u> could be **=2+2**, while a <u>function</u> might be **=Sum(A1:B1)**.

4.1. Insert references

It is important to know how to reuse data to avoid errors. This is facilitated by references.

4.1.1 Insert relative, absolute, and mixed references

There are three ways to reference other cells/ranges in Excel: relative, absolute and mixed. This section will explain each of these reference types.

Insert relative references

1. **Select** a cell
2. On the Formula Bar:
   a. Enter **=SUM(B3:E3)**
   b. Click the **Enter** icon (check mark symbol)
3. **Copy/Paste** the cell (not the formula bar contents) to the next row
4. **Observe** that the formula adjusted to be relative to the pasted cell

It should be noted that relative references are the default behavior; just entering a location (e.g. B3) will automatically be adjusted when copied to another location (e.g. B4).

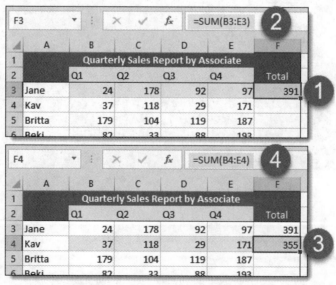

Relative references in formulas

Insert absolute references 🎥

1. **Select** a cell
2. On the Formula Bar:
    a. Enter **=SUM($B$3:$E$3)**
    b. Click the **Enter** icon (check mark symbol)
3. **Copy/Paste** the cell to the next row (and to the right, as a test)
4. **Observe** that the formula did not adjust, and the result is the same

Adding a $ (dollar sign) before each reference forces the formula to remain intact when it is copied. This can be helpful when a result (e.g. a total) needs to show up in multiple locations.

The only way an absolute reference will change is when a row or column is inserted. All references are adjusted to maintain data integrity.

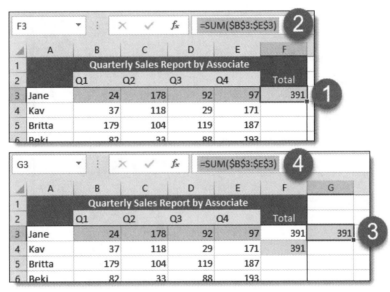

Absolute references in formulas

Insert mixed references

1. **Select** a cell
2. On the Formula Bar:
   a. Enter **=SUM(B3:E3)*$G$2**
   b. Click the **Enter** icon (check mark symbol)
3. **Copy/Paste** the cell to the next row
4. **Observe** that part of the formula adjusted, while the reference to the tax rate did not

Additionally, a single reference can be mixed. For example, B3 could be $B3 or B$3. This method is only used in rare cases.

The only way an absolute reference will change is when a row or column is inserted. All references are adjusted to maintain data integrity.

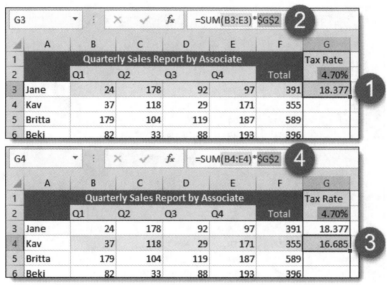

Using mixed references in formulas

4.1.2 Reference named ranges and named tables in formulas

Review the steps required to create and reference named ranges and tables.

Create and reference a named range in a formula 🎥

1. **Select** a range; e.g. **B3:E3**
2. In the **Name Box**, enter a name; e.g. **JaneQuarterly**
3. **Select** a blank cell; e.g. **F3**
4. In the Formula Bar:
   a. Enter **=SUM(JaneQuarterly)**
   b. Click the **Enter** icon (check mark symbol)

Observe that the total is the sum of all cells within the named range.

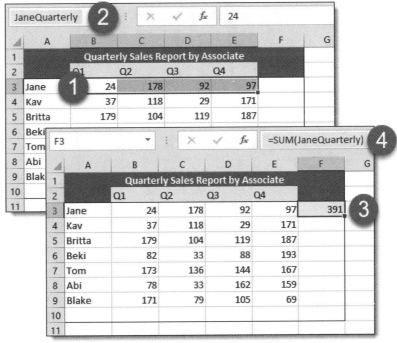

Create and reference a named range in a formula

Create and reference a named table in a formula

1. **Select** anywhere within a table
2. On the **Table Design** tab, enter a table name; e.g. AllQuarterlySales
3. **Select** a blank cell outside of the named table
4. In the Formula Bar:
   a. Enter **=SUM(AllQuarterlySales)**
   b. Click the **Enter** icon (check mark symbol)

Observe that the total is the sum of all cells within the named table. If the table includes row totals, those values will be included in this total, thus incorrectly doubling the result.

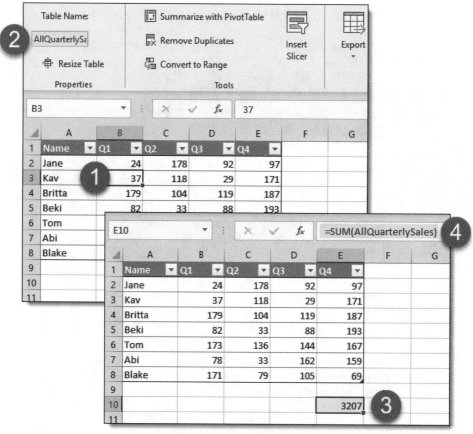

Create and reference a named table in a formula

4.2. Calculate and transform data

There are many ways in which data can be calculated and transformed in Excel. This section covers the basics every user should understand.

4.2.1 Perform calculations by using the AVERAGE(), MAX(), MIN(), and SUM() functions

Review a few of the most popular functions used to compute important results.

Perform calculations using the AVERAGE() function

1. **Select** a cell; e.g. F3
2. In the Formula Bar:
    a. Enter **=AVERAGE(B3:E3)**
    b. Click the **Enter** icon (check mark symbol)
3. **Observe** results; 37+118+29+171/4=88.75

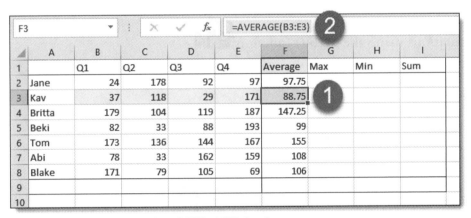

Perform calculations using the AVERAGE function

Additional examples:

The following examples also apply to Max, MIN and SUM.

- Named table: =AVERAGE(AllQuarterlySales)
- Named range: =AVERAGE(JaneQuarterly)
- Absolute reference: =AVERAGE($B$3:$E$3)
- Multiplication: =AVERAGE(B3:E3)*3
- Table reference   =AVERAGE(Table3[@[Column2]:[Column6]])

Perform calculations using the MAX() function

1. **Select** a cell; e.g. G3
2. In the Formula Bar:
    a. Enter **=MAX(B3:E3)**
    b. Click the **Enter** icon (check mark symbol)
3. **Observe** results; the largest number is 171

	A	B	C	D	E	F	G	H	I
		Q1	Q2	Q3	Q4	Average	Max	Min	Sum
2	Jane	24	178	92	97	97.75	178		
3	Kav	37	118	29	171	88.75	171		
4	Britta	179	104	119	187	147.25	187		
5	Beki	82	33	88	193	99	193		
6	Tom	173	136	144	167	155	173		
7	Abi	78	33	162	159	108	162		
8	Blake	171	79	105	69	106	171		
9									
10									

G3 | =MAX(B3:E3)

Perform calculations using the MAX function

Perform calculations using the MIN() function

1. **Select** a cell; e.g. H3
2. In the Formula Bar:
    a. Enter **=MIN(B3:E3)**
    b. Click the **Enter** icon (check mark symbol)
3. **Observe** results; the smallest number is 29

	A	B	C	D	E	F	G	H	I
		Q1	Q2	Q3	Q4	Average	Max	Min	Sum
2	Jane	24	178	92	97	97.75	178	24	
3	Kav	37	118	29	171	88.75	171	29	
4	Britta	179	104	119	187	147.25	187	104	
5	Beki	82	33	88	193	99	193	33	
6	Tom	173	136	144	167	155	173	136	
7	Abi	78	33	162	159	108	162	33	
8	Blake	171	79	105	69	106	171	69	
9									
10									

H3 | =MIN(B3:E3)

Perform calculations using the MIN function

Perform calculations using the SUM() function

1. **Select** a cell; e.g. I3
2. In the Formula Bar:
    a. Enter **=SUM(B3:E3)**
    b. Click the **Enter** icon (check mark symbol)
3. **Observe** results; 37+118+29+171=355

	A	B	C	D	E	F	G	H	I
1		Q1	Q2	Q3	Q4	Average	Max	Min	Sum
2	Jane	24	178	92	97	97.75	178	24	391
3	Kav	37	118	29	171	88.75	171	29	355
4	Britta	179	104	119	187	147.25	187	104	589
5	Beki	82	33	88	193	99	193	33	396
6	Tom	173	136	144	167	155	173	136	620
7	Abi	78	33	162	159	108	162	33	432
8	Blake	171	79	105	69	106	171	69	424
9									
10									

Perform calculations using the SUM function

Applying functions automatically

The following steps help to streamline the process of using functions in typical cases. These steps can be used for all the functions listed in the AutoSum list.

1. **Select** a range, including an adjacent blank cell
2. Click **Home → AutoSum arrow → Average**
3. **Observe** the results

If a single blank cell is selected, a range is automatically suggested but another can be selected. Click the Finish icon (check mark) on the formula bar to complete the process.

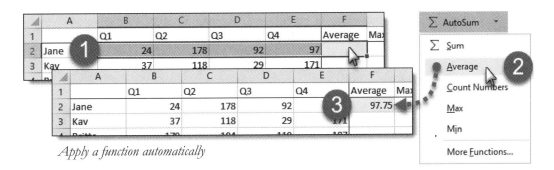

Apply a function automatically

4.2.2 Count cells by using the COUNT(), COUNTA(), and COUNTBLANK() functions

The following steps summarize the use of the three main COUNT functions in Excel.

Count cells using the COUNT() function

The COUNT() function counts the number of cells containing numbers.

1. **Select** a blank cell; e.g. F1
2. In the Formula Bar:
   a. Enter **=COUNT(B2:E8)**
   b. Click the **Enter** icon (check mark symbol)
3. **Observe** the results; there are 24 cells that contain numbers

The COUNT() function excludes blank cells and any cell containing a symbol or letter. The only exceptions are letters/symbols visible as a result of built-in cell formatting, e.g. currency showing a $ symbol; these <u>are</u> included in the count.

F1		f_x	=COUNT(B2:E8)		
	A	B	C	D	E	F
1		Q1	Q2	Q3	Q4	24
2	Jane	24	178	92	97	
3	Kav	37		29	171	
4	Britta	179	104	119	187	
5	Beki	82	33 Sabbatical		193	
6	Tom	173	136	144	167	
7	Abi		33	162	159	
8	Blake	171		105	69	

Count cells using the COUNT function

Count cells using the COUNTA() function

The COUNTA() function counts the number of non-blank cells.

1. **Select** a blank cell; e.g. F1
2. In the Formula Bar:
   a. Enter **=COUNTA(B2:E8)**
   b. Click the **Enter** icon (check mark symbol)
3. **Observe** the results; there are 25 non-blank cells

The COUNTA() function only excludes blank cells.

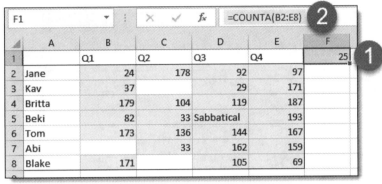

Count cells using the COUNTA function

Count cells using the COUNTBLANK() function

The COUNTBLANK() function counts the number of non-blank cells.

1. **Select** a blank cell; e.g. F1
2. In the Formula Bar:
   a. Enter **=COUNTBLANK(B2:E8)**
   b. Click the **Enter** icon (check mark symbol)
3. **Observe** the results; there are 3 blank cells

The COUNTBLANK() function only counts blank cells.

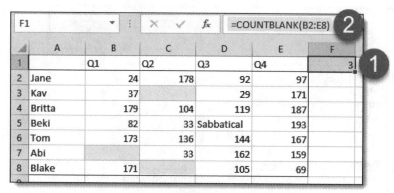

Count cells using the COUNTBLANK function

4.2.3 Perform conditional operations by using the IF() function

Using the IF function provides limitless opportunities to analyze data and then cause different things to happen. For example, we can determine IF each salesperson met the annual quota, returning a "Yes" or "No" depending on the answer to that question. Similarly, we can add a bonus to the total IF each salesperson sold more than a certain number of units.

IF function syntax:

IF --- Something is true --- then, do this --- otherwise, do that

For example:
In row 1, 4 is greater than 2; this is true, so do the first thing, i.e. return "1"
In row 2, 1 is not greater than 2; this is false, so do the second thing, i.e. return "0"

	A	B
1	4	1	=IF(A1>2, 1, 0)
2	1	0	=IF(A2>2, 1, 0)

Use IF() function to return text

1. **Select** a blank cell; e.g. F2
2. In the Formula Bar:
   a. Enter **=IF(SUM(B2:E2)>300,"Yes","No")**
   b. Click the **Enter** icon (check mark symbol)
3. **Observe** the results; only Kav did not meet the quota of 300

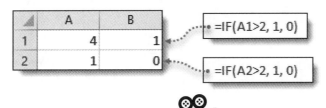

F2				f_x	=IF(SUM(B2:E2)>300,"Yes","No")
	A	B	C	D	E	F
1		Q1	Q2	Q3	Q4	Met Quota
2	Jane	24	178	92	97	Yes
3	Kav	37	15	29	171	No
4	Britta	179	104	119	187	Yes
5	Beki	82	33	0	193	Yes
6	Tom	173	136	144	167	Yes
7	Abi	0	33	162	159	Yes
8	Blake	171	44	105	69	Yes
9						

Use IF() function to return text

Use IF() function to return a calculated value 🎥

1. **Select** a blank cell; e.g. G2
2. In the Formula Bar:
   a. Enter **=IF(SUM(B2:E2)>400,SUM(B2:E2)+100,SUM(B2:E2))**
   b. Click the **Enter** icon (check mark symbol)
3. **Observe** the results; only Britta and Tom earned a $100 bonus

In this example, if the annual total is greater than 400, another 100 is added. Otherwise, the annual total is used. Notice the SUM function, SUM(B2:E2), is used three times: once in the IF "question" and then in a calculation if the "answer" is true/correct, and then again if the "answer" is false/incorrect.

	A	B	C	D	E	F	G	H
		Q1	Q2	Q3	Q4	Sub-Total	Add Bonus	
2	Jane	24	178	92	97	391	391	
3	Kav	37	15	29	171	252	252	
4	Britta	179	104	119	187	589	689	
5	Beki	82	33	0	193	308	308	
6	Tom	173	136	144	167	620	720	
7	Abi	0	33	162	159	354	354	
8	Blake	171	44	105	69	389	389	

G2 | fx =IF(SUM(B2:E2)>400,SUM(B2:E2)+100,SUM(B2:E2)

Use IF() function to return a calculated value

4.3. Format and Modify Text

Excel has several ways to dynamically modify and format text. That will be the topic of this section.

4.3.1 Format text by using RIGHT(), LEFT(), and MID() functions

This section will show three functions that can be used to extract a subset of data out of another cell.

Use RIGHT() function to return a truncated value

1. **Select** a blank cell; e.g. B2
2. In the Formula Bar:
   a. Enter =**RIGHT(A2, 3)**
   b. Click the **Enter** icon (check mark symbol)
3. **Observe** the results; only the three digits <u>from the right</u> are returned

In this example, for this formula to work, the department codes must all be exactly three digits. Once cell B2 is working, drag the grip down to copy the relative formula into the remaining cells in the column.

B2				=RIGHT(A2, 3)	
	A	B	C	D	E
1	Name - Dept Code	Dept Code	Q1	Q2	Q3
2	Jane - 001	001	24	178	
3	Kav - 001	001	37	15	
4	Britta - 001	001	179	104	
5	Beki - 002	002	82	33	
6	Tom - 002	002	173	136	
7	Abi - 100	100	0	33	
8	Blake - 100	100	171	44	

Use RIGHT() function to return truncated value

Use LEFT() function to return a truncated value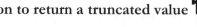

1. **Select** a blank cell; e.g. H2
2. In the Formula Bar:
   a. Enter =**LEFT(G2, 10)**
   b. Click the **Enter** icon (check mark symbol)
3. **Observe** the results; only the ten digits <u>from the left</u> are returned

In this example, for this formula to work, the date must consistently be entered in the same 10 digit format. Once cell H2 is working, drag the grip down to copy the relative formula into the remaining cells in the column.

H2		fx	=LEFT(G2, 10)	
	E	F	G	H	I
1	Q3	Q4	Date/Time Submitted	Date Only	
2	92	97	12/28/2008; 8:00am	12/28/2008	
3	29	171	12/28/2008; 10:35am	12/28/2008	
4	119	187	12/22/2008; 3:00pm	12/22/2008	
5	0	193	12/10/2008; 11:00am	12/10/2008	
6	144	167	12/01/2008; 7:00am	12/01/2008	
7	162	159	12/17/2008; 4:20pm	12/17/2008	
8	105	69	12/13/2010; 2:00pm	12/13/2010	

Use LEFT() function to return truncated value

Use MID() function to return a truncated value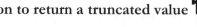

1. **Select** a blank cell; e.g. K2
2. In the Formula Bar:
   a. Enter =**MID(J2, 5, 3)**
   b. Click the **Enter** icon (check mark symbol)
3. **Observe** the results; only the 3 digits <u>5 digits over from the left</u> are returned

In this example, the desired numbers must always have four digits on the left.

K2		fx	=MID(J2, 5,3)	
	I	J	K	L	M
1		Serial Number	Batch ID		
2		100-108-12004	108		
3		100-108-12204	108		
4		100-122-12305	122		
5		200-122-13009	122		
6		200-143-13107	143		
7		300-143-13008	143		
8		300-143-18012	143		

Use MID() function to return truncated value

4.3.2 Format text by using UPPER(), LOWER(), and LEN() functions

In this section you will review the steps required to change the case of text and count the number of characters contained within a given cell.

Use UPPER() function to change text

1. **Select** a blank cell; e.g. B2
2. In the Formula Bar:
    a. Enter **=Upper(A2)**
    b. Click the **Enter** icon (check mark symbol)
3. **Observe** the results; all letters are now uppercase

If the text in A2 is changed, the results in B2 will automatically update.

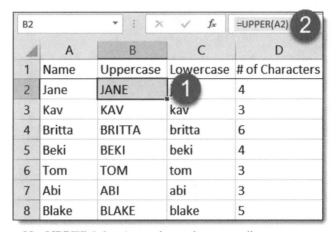

Use UPPER() function to change the text to all uppercase

Use LOWER() function to change text

1. **Select** a blank cell; e.g. C2
2. In the Formula Bar:
   a. Enter **=LOWER(A2)**
   b. Click the **Enter** icon (check mark symbol)
3. **Observe** the results; all letters are now lowercase

If the text in A2 is changed, the results in C2 will automatically update.

| C2 | | fx | =LOWER(A2) | 2 |

	A	B	C	D
1	Name	Uppercase	Lowercase	# of Characters
2	Jane	JANE	jane	
3	Kav	KAV	kav	3
4	Britta	BRITTA	britta	6
5	Beki	BEKI	beki	4
6	Tom	TOM	tom	3
7	Abi	ABI	abi	3
8	Blake	BLAKE	blake	5

Use LOWER() function to change the text to all lowercase

Use LEN() function to count characters

1. **Select** a blank cell; e.g. D2
2. In the Formula Bar:
   a. Enter **=LEN(A2)**
   b. Click the **Enter** icon (check mark symbol)
3. **Observe** the results; there are 4 characters in cell A2

If the text in A2 is changed, the results in D2 will automatically update.

| D2 | | fx | =LEN(A2) | 2 |

	A	B	C	D
1	Name	Uppercase	Lowercase	# of Characters
2	Jane	JANE	jane	4
3	Kav	KAV	kav	3
4	Britta	BRITTA	britta	6
5	Beki	BEKI	beki	4
6	Tom	TOM	tom	3
7	Abi	ABI	abi	3
8	Blake	BLAKE	blake	5

Use LEN() function to count the number of characters

4.3.3 Format text by using the CONCAT() and TEXTJOIN() functions

This section reviews ways to combine data for a more cohesive and easier to read look.

Use CONCAT() function to combine text in multiple cells

1. **Select** a blank cell; e.g. C2
2. In the Formula Bar:
   a. Enter **=CONCAT(A2, " ",B2)**
   b. Click the **Enter** icon (check mark symbol)
3. **Observe** the results; first and last name combined into new cell separated by a space

> Column A and Column B could now be hidden as the information is redundant.

To insert additional characters, include a new item in the function separated by quotation marks. This is how the space was introduced in this example.

	A	B	C	D		C	
			=CONCAT(A2, " ",B2)			=CONCAT(A2, " ",B2,", Esq")	
1	First Name	Last Name	Full Name	Delimited		Full Name	De
2	Jane	Anderson	Jane Anderson	Anderson		Jane Anderson, Esq	
3	Kav	Asadi	Kav Asadi	Kav - Asadi		Kav Asadi, Esq	Ka
4	Britta	Elder	Britta Elder	Britta - Elder		Britta Elder, Esq	Br
5	Beki	Bentley	Beki Bentley	Beki - Bentley		Beki Bentley, Esq	Be
6	Tom	Mohan	Tom Mohan	Tom - Mohan		Tom Mohan, Esq	To
7	Abi	Strauser	Abi Strauser	Abi - Strauser		Abi Strauser, Esq	Ab
8	Blake	Volz	Blake Volz	Blake - Volz		Blake Volz, Esq	Bla

Use the CONCAT() function to combine text in multiple cells

To expand on this idea, the image above includes an additional example. If all the names in the list are attorneys, add another item to the function to include the suffix Esq.

The formula would look like this: **=CONCAT(A2, " ",B2,", Esq")**

Another example might be creating an email address.

> For example: =CONCAT(A1,"@ACME.com")
> The result would be: **Jane@ACME.com**

To make the email address all lower case, use a nested function.

> For example: =LOWER(=CONCAT(A1,"@ACME.com"))
> The result would be: **jane@acme.com**

Use TEXTJOIN() function to combine text and separate

1. **Select** a blank cell; e.g. D2
2. In the Formula Bar:
   a. Enter **=TEXTJOIN(" - ",TRUE,A2:B2)**
   b. Click the **Enter** icon (check mark symbol)
3. **Observe** the results; the two names are combined and separated by a dash with spaces.

Notice the delimiter is a Space, a Dash and then another Space all enclosed in quotation marks.

D2			fx	=TEXTJOIN(" - ",TRUE,A2:B2) ②
	A	**B**	**C**	**D**
1	First Name	Last Name	Full Name	Delimited
2	Jane	Anderson	Jane Anderson	Jane - Anderson ①
3	Kav	Asadi	Kav Asadi	Kav - Asadi
4	Britta	Elder	Britta Elder	Britta - Elder
5	Beki	Bentley	Beki Bentley	Beki - Bentley
6	Tom	Mohan	Tom Mohan	Tom - Mohan
7	Abi	Strauser	Abi Strauser	Abi - Strauser
8	Blake	Volz	Blake Volz	Blake - Volz

Use the TEXTJOIN() function to combine and separate text

Alternately, rather than specifying individual cells, a range may be used, as shown in the following image.

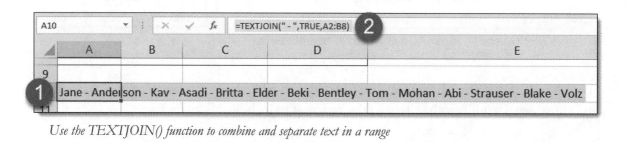

Use the TEXTJOIN() function to combine and separate text in a range

4.0 Practice Tasks

Try the topics covered in this chapter to make sure you understand the concepts. These tasks are sequential and should be completed in the same Excel workbook unless noted otherwise. Saving the results is optional, unless assigned by an instructor.

Task 4.1:

✓ Open Monthly Sales Report by Associate.xlsx, and then create a **named range** called FullTable for range B3:M9. Finally, **create a formula** that Sums the total of FullTable.

Task 4.2

✓ In column N, **sum the total** annual sales for each associate.

Task 4.3

✓ Manually enter the value 0.75 in cell O1. Create a formula that multiplies the total from Column N with the fixed value in O1, using **Mixed and Absolute references**. Use AutoFill to copy the formula to all rows (i.e. for each associate).

Task 4.4:

✓ Create a formula to calculate the **Average**, **Maximum** and **Minimum** values for each associate. Add a header in row 2 as shown below.

Task 4.5:

✓ Create a formula, in column S, to change each associate's name to **all uppercase**.

L	M	N	O	P	Q	R	S	T
			0.75					
	Dec			Avg	Max	Min		
92	97	1218	913.5	101.5	190	21	JANE	
29	171	883	662.25	73.58333	198	15	KAV	
119	187	1599	1199.25	133.25	187	50	BRITTA	
88	193	1055	791.25	87.91667	193	17	BEKI	
144	167	1444	1083	120.3333	191	45	TOM	
162	159	1219	914.25	101.5833	163	28	ABI	
105	69	1242	931.5	103.5	171	12	BLAKE	
		8660						

Notes:

5.0 Manage Charts

Introduction

This chapter reviews the steps required to format charts in Microsoft Excel.

5.1.   Create Charts

This section reviews exam outcomes related to managing charts in Excel.

5.1.1 Create charts

Excel charts bring data visualization to a whole new level. Review the steps to create one here.

Create Recommended Charts

1. **Select** a table by clicking anywhere on it.
2. Click **Insert → Recommended Charts** *FYI: notice simple Pie and Bar chart commands also*
   a. Notice additional chart options: 2-D Pie Chart, 3-D Column chart and more.
3. Insert Charts dialog: Select an option and click **OK** (see image on next page).
4. **Review** the results.

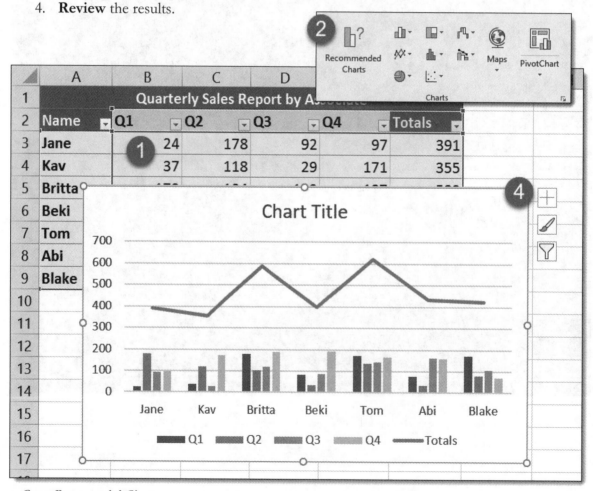

The chart can be dragged to any location in the worksheet. The **chart title** can be changed by clicking on it multiple times until in edit mode.

To delete a chart, simple select it and press the **Delete** key.

When the chart is selected, three icons appear in the upper right. Each of these icons is described below, along with an image showing their options:

A. **Chart Elements**
   Add, Change or review chart elements, e.g.

B. **Chart Styles**
   Manage the style and color scheme; notice the two tabs at the top of the panel.

C. **Chart Filters**
   Control what data appears in the chart.

Select a chart template to create

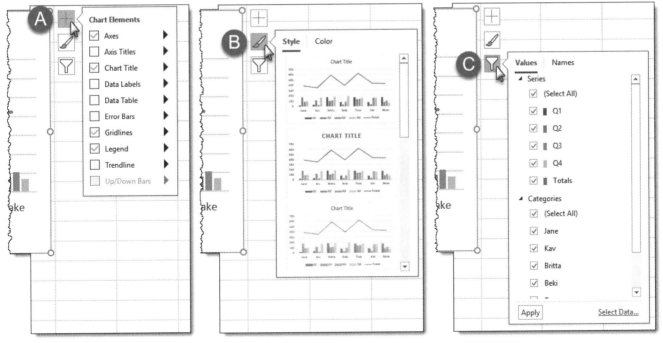

Chart editing options

5.1.2 Create chart sheets

Charts can be moved to their own worksheet, rather than floating on top of the one it was created on. This is a good way to organize data and make charts easier to find.

Create Chart Sheets

1. **Select** a table by clicking anywhere on it.
2. On the Ribbon, click **Chart Design → Move Chart**
3. Move Chart dialog:
    a. Select **New Sheet**
    b. **Enter a name**; e.g. Quarterly Sales Chart
    c. Click **OK**.
4. **Review** results.

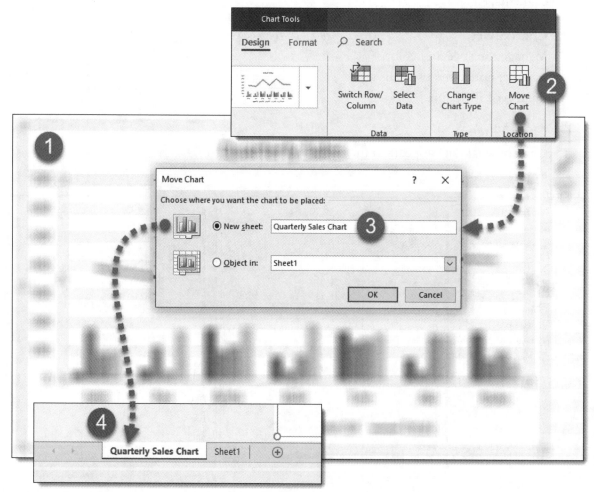

Create Chart Sheets

5.2.   Modify Charts

There are several ways in which charts may be modified. That is the focus of this section.

5.2.1 Add data series to charts

When new rows are added to a range or table, the chart will often update automatically to include the new information. However, that is not always the case. Here is an example of a new row, a new associate starting in the middle of Q3, and how this data is manually made to be included in the chart that already exists.

Add a new row to a chart

1. **Select** a chart by clicking anywhere on it; notice the defining cells highlight.
2. Click and drag the bottom edge of the highlighted range down to include the new row.
3. **Review** results; the chart includes the new associate and her sales.

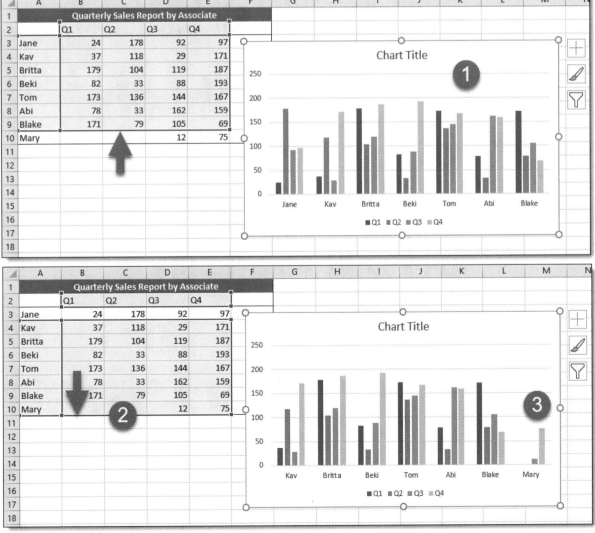

Add a data series to a chart

5.2.2 Switch between rows and columns in source data

It is sometimes helpful to switch the way rows and columns are referenced in a chart, such that data is compared by columns rather than by rows. Here's how that is done:

Switch between rows and columns in a chart

1. **Select** a chart by clicking anywhere on it.
2. Click **Chart Design → Switch Row/Column**.
3. **Review** the results.

The sales are now being compared by quarter, rather than by individual.

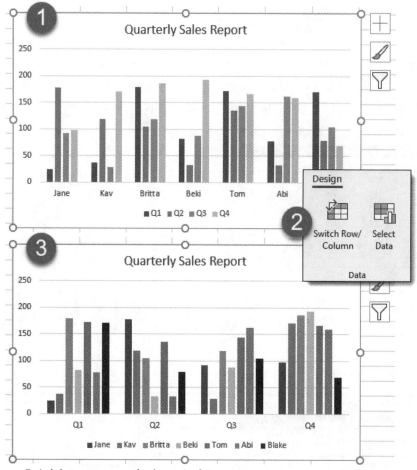

Switch between rows and columns in data source

5.2.3 Add and modify chart elements

This section reviews how chart elements are added and modified.

Review chart element toggles

1. **Select** a chart by clicking anywhere on it.
2. Click the chart elements icon.
3. **Toggle** options or click **arrow** for more options.
4. **Review** the results.

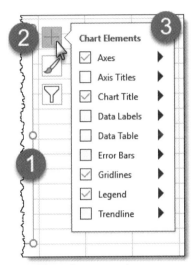

Chart elements toggles

A few examples of how these toggles affect the chart are shown to the right.

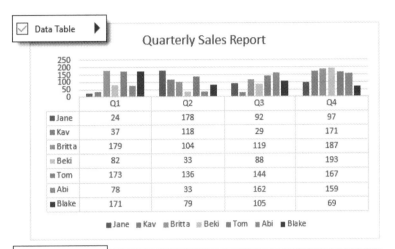

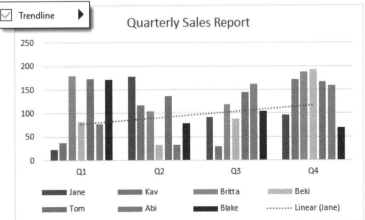

Compare chart element toggle options

In addition to the dynamic toggles associated with selecting a chart, there are many more formatting options. Here is how they are accessed:

Modify chart format properties

1. **Select** a chart by clicking anywhere on it and **Right-click** on any part; e.g. the legend.
2. Click **Format Legend**; the name varies based on the element selected.
3. Make changes to the format properties.
4. **Review** the results.

To access additional properties, the following two items correspond to the lettered items in the image below.

A. **Chart options:** access properties for all elements within the selected chart; e.g. chart title, plot area, and each series item in the chart.

B. **Graphic tabs:** related formatting options are grouped into tabs; e.g. Fill & Line, Effects and Legend Options.

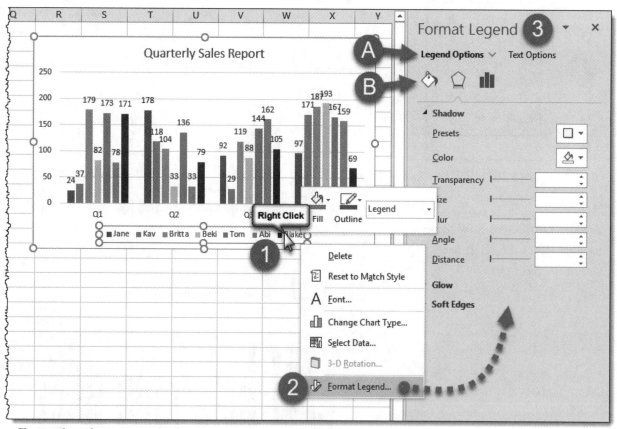

Format chart elements

5.3. Format charts

This section reviews ways in which charts might have their formatting modified.

5.3.1 Apply chart layouts

Chart layouts, or Quick Layouts, apply several chart element format changes at once.

Apply Quick Layouts

1. **Select** a chart by clicking anywhere on it.
2. On the Ribbon, click **Design → Quick Layout**.
3. In the expanded list:
   a. Hover over an option to see a preview and a formatting list (tooltip).
   b. Click to select an option.
4. **Review** the results.

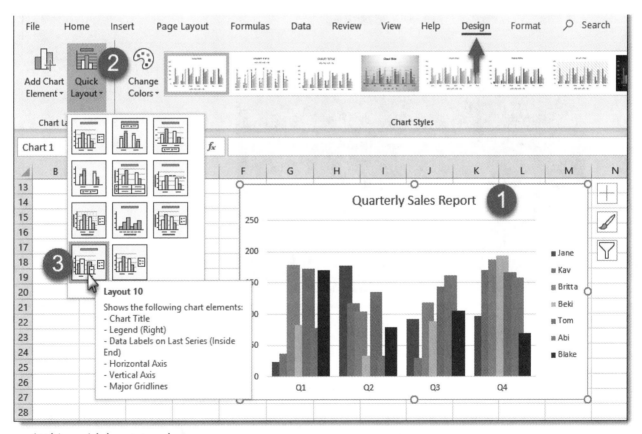

Applying quick layouts to a chart

5.3.2 Apply chart styles

Styles are more than basic layout options. They facilitate dramatic graphical representation of data. A style is selected/applied when the chart is originally created. However, it can easily be changed at any time. Here's how:

Apply Chart Styles

1. **Select** a chart by clicking anywhere on it.
2. Click **Design** → **Styles**.
   a. Hover over an option to see a preview.
   b. Click to select an option.
3. **Review** the results.

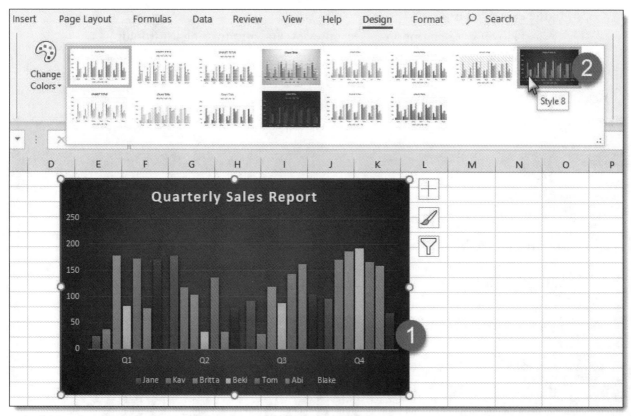

Applying a chart style to an existing chart

Another element of a chart style is the **Color Palette**. This may be changed by selecting the chart and then picking a color palette option from the list (see image to the right). Notice, when hovering over a row, the palette name appears. Finally, take note of the two sections: **Colorful** and **Monochromatic**.

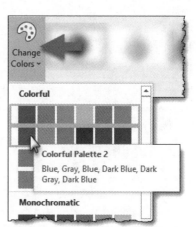

5.3.3 Add alternative text to charts for accessibility

Excel provides a method to store text which can be read out loud to individuals with visual impairments. As shown in the screen capture below, 1 – 2 sentences are recommended.

Adding alternative text to charts

1. **Select** a chart by clicking anywhere on it.
2. Click **Format → Alt Text**.
3. Enter 1 – 2 sentences in the Alt Text panel that describes your chart to someone who is blind.

Adding alternative text to charts

5.0 Practice Tasks

Try the topics covered in this chapter to make sure you understand the concepts. These tasks are sequential and should be completed in the same Excel workbook unless noted otherwise. Saving the results is optional, unless assigned by an instructor.

Task 5.1:

- ✓ Open Quarterly Sales by Associate.xlsx, and then select the range A2:E9. Create a **2D Line** chart.

Task 5.2

- ✓ Move the new chart to a **new sheet** (aka worksheet).

Task 5.3

- ✓ Add **Data Labels** via Chart Elements.

Task 5.4:

- ✓ Switch the **Rows and Columns** in the chart.

Task 5.5:

- ✓ Add the following **Alt Text** to the chart: Quarterly Sales Report by Associate.

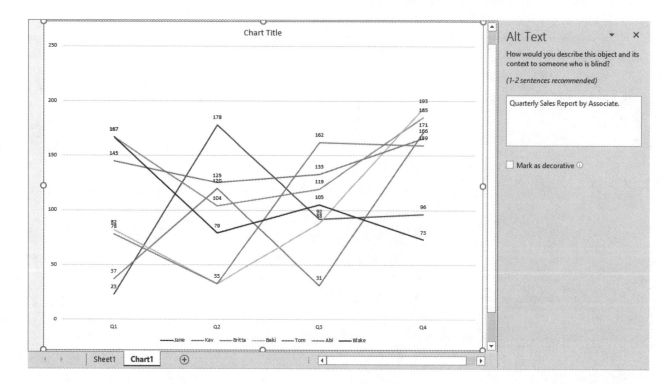

6.0 Practice Exam

Introduction

This chapter will highlight the practice exam software provided with this book, including accessing the exam, Installing, required files, user interface and how to interpret the results. Taking this practice exam, after studying this book, will help ensure a successful result when taking the actual Microsoft Office Specialist Certified Associate exam at a test center.

The practice exam questions are similar, not identical, to the actual exam.

Important Things to Know

Here are a few big picture things you should keep in mind:

- **Practice Exam – First Steps**
  - The practice exam, that comes with this book, is taken on **your own computer**
  - You need to have **Excel installed** and ready to use during the practice exam
  - You must download the practice exam software from SDC Publications
    - See inside-front cover of this book for access instructions
  - **Required Excel files** for the practice exam
    - Files downloaded with practice exam software
    - Locate files before starting practice test
  - Note which questions you got wrong, and study those topics

- **Practice Exam - Details**
  - Questions: 35
  - Timed: 50 minutes
  - Passing: 70%
  - Results: Presented upon completion

This practice exam can be taken multiple times. But it is recommended that you finish studying this book before taking the practice exam. There are only 35 questions total, so you don't want to get to a point where all the questions, and their answers, have been memorized. This will not help with the actual exam as they are not the same questions.

This practice exam can be taken multiple times.

Practice Exam Overview

The **practice exam** included with this book can be downloaded from the publisher's website using the **access code** found on the inside-front cover. This is a good way to check your skills prior to taking the official exam, as the intent is to offer similar types of questions in roughly the same format as the formal exam. This practice exam is taken at home, work or school, on your own computer. You must have Excel installed to successfully answer the in-application questions.

This is a test drive for the exam process:

- Understanding the test software
- How to mark and return to questions
- Exam question format
- Live in-application steps
- How the results are presented at the exam conclusion

Here is a sample of what the practice exam looks like... note that Excel is automatically opened and positioned directly above the practice exam user interface.

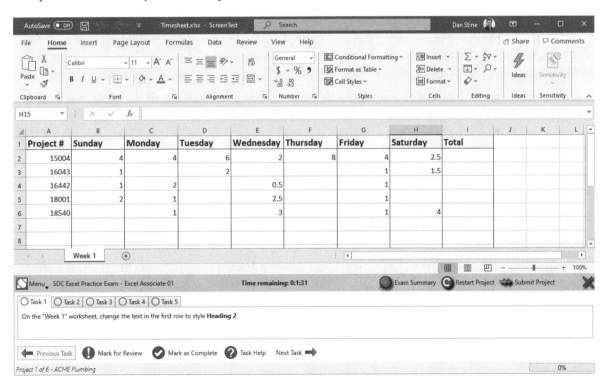

Sample question from included practice exam

Having taken the practice exam can remove some anxiety one may have going into an exam that may positively impact a career search or advancement.

Download and Install the Practice Exam

Follow the instructions on the inside-front cover of this book, using the provided access code to download the practice exam. Once the ZIP file is downloaded you must extract the files into a folder that you create.

Download steps:

- Create a folder on your desktop or C drive, such as **C:\MOS Installer**
- Double-click on the downloaded ZIP file
- Copy all the folders/files from the ZIP file to the newly created folder

To install the practice Exam software on your PC-based computer simply double-click the Setup.exe file in the newly created folder. Follow the prompts on screen to complete the installation. Once complete, the folder just created, and its contents, may be deleted.

Required Excel Files

The installed practice exam software includes several required Excel files to be used during the exam. For the most part, the software will open the files when they are needed. There are, however, a few questions that require a file be selected and imported. In those cases, the current working folder is changed so the file should be directly accessible when trying to access it.

For the practice exam, the required files are installed automatically.

Starting the Exam

From the Windows Start menu, click the **SDC Practice Exam** icon to start the practice exam. If you have purchased and installed more than one SDC practice exam, select the desired practice exam from the list that appears. At this point, the practice exam opens with the timer running. Excel is also opened, along with the required workbook.

Note the following formatting conventions used in the exam questions:
- **Bold text** is used to indicate file or folder names as well as setting names.
- Clicking on underlined text copies it to the clipboard. Use **Ctrl + V** to paste into Excel to avoid typing errors.
- Text in "quotation marks" represents existing text within the workbook.

Practice Exam User Interface (UI)

The following image, and subsequent list, highlight the features of the practice exam's user interface.

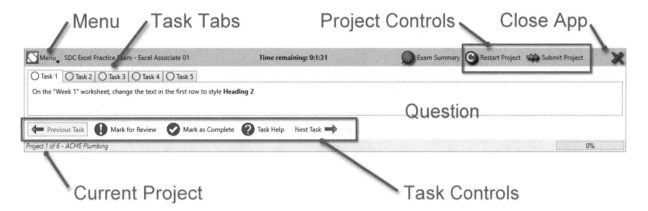

User Interface details:

- **Menu:** (drop-down list)
    - o **Float Application Window –** Use to reposition practice exam on the desktop
    - o **Dock to Desktop Bottom –** Default option, exam fixed to bottom of screen
    - o **Exam Summary –** Review marked questions
    - o **Finish Exam –** Grade the exam
    - o **About –** Exam software version information
    - o **Close –** Closes the Practice Exam and Excel
- **Task Tabs:** Each tab contains a question for the current project and may be marked for review or as completed. Click a tab to view its question or use Previous/Next buttons.
- **Time Remaining:** Time remaining for the 50 minute timed exam
- **Exam Summary:** Review marked questions and return to previous project/question
- **Project Controls:**
    - o **Reset Project –** Discards all changes made to the current workbook
    - o **Submit Project –** Advance to next project or exam completion on last project
- **Close App:** Closes the Practice Exam and Excel
- **Current Project:** Current project name and number listed for reference
- **Task Controls:** *for the current project…*
    - o **Previous Task –** View the previous task/question
    - o **Mark for Review –** *Optional:* When unsure of the answer, mark task for review
    - o **Mark as Complete –** *Optional:* When confident, mark the task as complete
    - o **Task Help –** *Optional:* Reveal steps required to achieve a correct answer
    - o **Next Task –** Advance to the next task/question

When unsure of the correct answer, after multiple attempts, click the **Task Help** button to reveal the steps required to answer the current question. The image below shows an example.

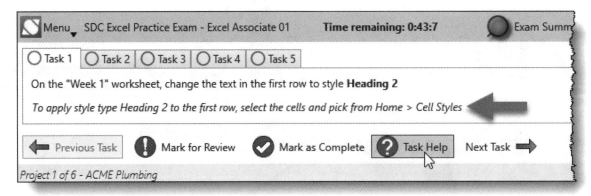

Practice exam - Help example

The following image shows an example of tasks marked for review and as complete. This is optional, and just meant as a way of tracking one's progress. It is possible to advance to the next project without marking any tasks.

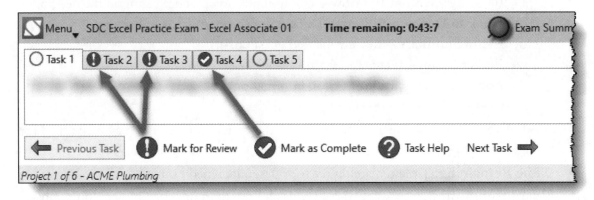

Practice exam – Marked task exams

Practice Exam Results

When you complete the practice exam, you will find out if you passed or failed. Be sure to note which questions were answered incorrectly and review those related sections in the book.

Conclusion

As with any formal exam, the more you practice the more likely you are to have successful results. So, be sure to take the time to download the provided practice exam and give it a try before you head off to the testing facility and take the actual exam.

Good luck!

Microsoft Office Specialist – Excel Associate 365/2019 - Exam Preparation
Exam Day Study Guide

Remembering where the right tools and commands are is half the battle. Leading up to exam day, use this handy reference sheet to firm up your knowledge of important topics that will help you pass the Excel exam.

Reminders
- ✓ Bring **photo ID**
- ✓ Know your Certiport **username** and **password**
- ✓ Bring exam **payment confirmation**
- ✓ Know where the testing center is; e.g. have the **building address**

Right-click
- ✓ Links (aka Hyperlink)
- ✓ Insert (Row, Column, Cell)
- ✓ Delete (Row, Column, Cell)
- ✓ Row Height
- ✓ Column Width
- ✓ Format Cells

Commands
- ⬇ **Home** tab
  - ✓ Format Painter
  - ✓ Wrap Text
  - ✓ Merge Cells
  - ✓ Number format
  - ✓ Conditional Formatting
  - ✓ Format as Table (i.e. create table)
  - ✓ Cell Styles
  - ✓ Clear: *All, Formats, Contents, Comments & Notes, Hyperlink*
  - ✓ Sort and Filter → Custom Sort
  - ✓ Find & Select: *Find, Replace, Formulas, Notes, Conditional Formatting, etc.*
- ⬇ **Insert** tab
  - ✓ Charts: Recommended, Bar, Pie, etc.
  - ✓ Sparklines: Line, Column, Win/Loss
- ⬇ **Page Layout** tab
  - ✓ Page Setup dialog (small icon in lower right): *Manage Headers/Footers*
  - ✓ Set Print Area
- ⬇ **Formulas** tab
  - ✓ Show Formulas
- ⬇ **Data** tab
  - ✓ Get Data → From File…
- ⬇ **View** tab
  - ✓ Freeze Panes: *Top Row, First Column*

Tips
- ✓ Click the **worksheet tab** listed in the question, to be sure it is active
- ✓ In the Name Box, always type the **cell location** or **range** given in the question
- ✓ Watch the **tooltip** in the formula bar to help get the syntax right
- ✓ Click underlined text to copy it to the clipboard
- ✓ Accept all defaults unless otherwise instructed

Functions and Formulas
Some formula names are more abstract, and others are abbreviated.

Name:	Example:
✓ AVERAGE	=AVERAGE(B3:E3)
✓ MAX	=MAX(B3:E3)
✓ MIN	=MIN(B3:E3)
✓ SUM	=SUM(B3:E3)
✓ COUNT	=COUNT(B2:E8)
✓ COUNTA	=COUNTA(B2:E8)
✓ COUNTBLANK	=COUNTBLANK(B2:E8)
✓ IF	=IF(SUM(B2:E2)>300,"Yes","No")
✓ RIGHT	=RIGHT(A2, 3)
✓ LEFT	=LEFT(G2, 10)
✓ MID	=MID(J2, 5, 3)
✓ UPPER	=UPPER(A2)
✓ LOWER	=LOWER(A2)
✓ LEN	=LEN(A2)
✓ CONCAT	=CONCAT(A2, " ",B2)
✓ TEXTJOIN	=TEXTJOIN(" - ",TRUE,A2:B2)
✓ Absolute	=SUM(B3:E3)
✓ Relative	=SUM($B$3:$E$3)
✓ Mixed	=SUM(B3:E3)*$G$2
✓ Named rage	=SUM(**JaneQuarterly**)

Notes:

Index

SDC PUBLICATIONS **Which two file types can Excel import (ones required on exam)?**	Page Layout, Page Setup (→ dialog launcher icon)
SDC PUBLICATIONS **Which ribbon tab do you use to import .txt or .cvs files?**	Sheet
SDC PUBLICATIONS **The Find & Select command is on which tab?**	1. Right-click cell 2. Column Width command 3. Enter value
SDC PUBLICATIONS **In the Name Box, type this to select range A4 through A8**	Page Setup
SDC PUBLICATIONS **The Link command can be found on which ribbon tab?**	1. Right click command 2. Add to Quick Access Toolbar command

SDC PUBLICATIONS	
Name the Ribbon and Panel location for Page Setup	.txt and .cvs

SDC PUBLICATIONS	
Name the tab, in the Page Setup dialog, used to define 'Print Area'	Data (→ Get Data → From File → From Text/CSV)

SDC PUBLICATIONS	
List the three steps to change row height	Home

SDC PUBLICATIONS	
Name the dialog used to create/modify headers and footers	A4:A8

SDC PUBLICATIONS	
List steps required to add a command to Quick Access Toolbar	Insert

List three workbook view options

1. Select range
2. Page Layout tab
3. Print Area → Set Print Area

Name the command on View tab to prevent the first row or column from moving

1. File tab
2. Save-As
3. Select file type
4. Save command

Name the command used to open another view of the same workbook

1. File tab
2. Print
3. Configure settings/options
4. Print

Name the location to view/edit workbook properties

File (tab) → Info

Which ribbon tab is the Show Formulas command located on?

Page Layout (tab) → Page Setup (dialog launcher) → Header/Footer (tab) → Custom Header/Footer (button)

Page 2b Questions and answers do not align on this page. Print Page 2a on the backside of this page and cut out flashcards.

SDC PUBLICATIONS **List the steps to define Print Area**	1. Normal 2. Page Break View 3. Page Layout
SDC PUBLICATIONS **List the steps to save workbook in alterative formats**	Freeze Pages
SDC PUBLICATIONS **List the steps to print / configure print settings**	New Window
SDC PUBLICATIONS **Name the location of 'Check for Issues' options**	File (tab) → Info → Properties
SDC PUBLICATIONS **Name the location to customize headers and footers**	Formulas

SDC PUBLICATIONS

Name the command to copy just the value from a cell with a formula

1. Select cell or range
2. Home → Wrap Text

SDC PUBLICATIONS

Describe how to Auto Fill a cell's contents to adjacent cells

1. Select desired cell
2. Home → Number list

SDC PUBLICATIONS

What is the pattern (or syntax) required to Auto Fill only even numbers starting at 8?

Format Cells dialog

SDC PUBLICATIONS

List the steps required to merge two or more cells

1. Select cell or range
2. Home → Cell Styles
3. Select from options

SDC PUBLICATIONS

Name the command to transfer visual appearance settings from one cell to another

Clear Formats

SDC PUBLICATIONS **List the steps required to wrap text within a cell**	Paste Special
SDC PUBLICATIONS **List the steps required to set a cell as currency**	1. Select desired cell 2. Drag grip in lower right corner
SDC PUBLICATIONS **Name the dialog to change number precision**	8,10
SDC PUBLICATIONS **List the steps to apply cell styles**	1. Select desired cell 2. Home → Merge & Center
SDC PUBLICATIONS **Name the command to remove all cell formatting**	Format Painter

SDC PUBLICATIONS

Selecting a range and typing a name in the Name Box creates a?

1. Select upper left cell in range
2. Home → Format as Table
3. Select or type range
4. Click OK

SDC PUBLICATIONS

List the steps to change the name of a table

Convert to Range

SDC PUBLICATIONS

On which ribbon tab can the Sparkline commands be found?

Table Design

SDC PUBLICATIONS

List the steps required to apply conditional formatting

Right-click

SDC PUBLICATIONS

Name the command used to remove conditional formatting from current selection

Table Style Options

SDC PUBLICATIONS List the steps to create a table	Named Range
SDC PUBLICATIONS Name the command used to remove a table, but leave the data	1. Click within a table 2. Design tab (only appears after selecting within a table) 3. Edit name in Properties panel
SDC PUBLICATIONS Name the contextual tab with table style options	Insert
SDC PUBLICATIONS Do this within a table to access the 'Insert' row or column commands	1. Select desired cell or range 2. Home → Conditional Formatting 3. Select from options
SDC PUBLICATIONS This panel, on the Table Design contextual tab, has the Total Row toggle	Clear Rules from Selected Cells

SDC PUBLICATIONS **This 'table style option' adds a row at the bottom of the schedule with a calculated grand total**	=SUM(B3:$E$3) *or* =SUM($B$3:E3) *or* =SUM(B3:$E3) *etc.*
SDC PUBLICATIONS **What does the down-arrow next to each column header in a table do?**	=SUM(JaneQuarterly)
SDC PUBLICATIONS **Name the ribbon location, and command, to sort data by multiple columns**	=Average(A3:E3)
SDC PUBLICATIONS **=SUM(B3:E3) is Absolute or Relative?**	MAX *(not maximum)*
SDC PUBLICATIONS **=SUM($B$3:$E$3) is Absolute or Relative?**	MIN *(not minimum)*

SDC PUBLICATIONS	
Change this to a mixed reference =SUM(B3:E3)	Total Row
SDC PUBLICATIONS	
Write a formula to sum the named range JaneQuarterly	Filters and Sorts data
SDC PUBLICATIONS	
Write a formula to find the average value for numbers in the third row in columns A through E	Home → Sort & Filter → Custom Sort
SDC PUBLICATIONS	
Name the function to find maximum value	Relative
SDC PUBLICATIONS	
Name the function to find Minimum value	Absolute

SDC PUBLICATIONS	
Name the function that reports how many cells contain a number	108

SDC PUBLICATIONS	
Name the function reporting how many cells contain a number or text	UPPER()

SDC PUBLICATIONS	
Name the function reporting the number of blank cells	LOWER()

SDC PUBLICATIONS	
A1 = 4 =IF(A1>2, 1, 0) Result = ?	LEN()

SDC PUBLICATIONS	
A2 = Jane – 001 =RIGHT(A2, 3) Result = ?	CONCANT()

SDC PUBLICATIONS J2 = 100-108-12004 =MID(J2, 5, 3) Result = ?	Count()
SDC PUBLICATIONS **Name the function to change text to all uppercase characters**	COUNTA()
SDC PUBLICATIONS **Name the function to change text to all lowercase characters**	COUNTBLANK()
SDC PUBLICATIONS **Name the function reporting the number of characters in a cell**	1
SDC PUBLICATIONS **Name the function used to merge the contents of two cells**	001

SDC PUBLICATIONS **Name the ribbon tab with tools to create a chart**	Right-click
SDC PUBLICATIONS **Name the command used to create a chart sheet from an existing chart**	Chart Layouts (Chart Design ribbon tab)
SDC PUBLICATIONS **Select this first, to add a new row to a chart**	Chart Styles
SDC PUBLICATIONS **Name the contextual ribbon tab with Switch Row/Column command**	File
SDC PUBLICATIONS **List the steps to modify chart elements**	50 minutes

SDC PUBLICATIONS Modify chart properties via?	Insert
SDC PUBLICATIONS Apply several chart element format changes at once with?	Move Chart
SDC PUBLICATIONS Apply this to a chart to change graphical appearance?	Select the chat (which then highlights the range)
SDC PUBLICATIONS The Print command is found on the following ribbon tab?	Chart Design
SDC PUBLICATIONS Time allotted for the exam?	1. Select chart 2. Click "+" icon in upper right 3. Toggle options in list